Prevention and Treatment of Delayed Ischaemic Dysfunction in Patients with Subarachnoid Haemorrhage

An Update

Edited by

H.-J. Reulen and J. Philippon

Acta Neurochirurgica
Supplementum 45

Springer-Verlag Wien New York

Professor Dr. Hans-Jürgen Reulen
Department of Neurosurgery, University Hospital, Inselspital, Bern, Switzerland

Professor Dr. Jacques Philippon
Department of Neurosurgery, C.H.U. Pitié-Salpêtrière, Paris, France

With 16 Figures

ISSN 0065-1419

ISBN-13:978-3-211-82096-4   e-ISBN-13:978-3-7091-9014-2
DOI: 10.1007/978-3-7091-9014-2

Preface

During the last meeting of the European Association of Neurosurgical Societies (Barcelona 7–16 September 1987), a Symposium was devoted to the use of calcium antagonists in cerebral vasospasm. As shown by its title "Prevention and treatment of delayed ischaemic dysfunction in patients with subarachnoid haemorrhage: an update", papers presented at the Symposium covered a broad spectrum from some basic data on pathophysiology of subarachnoid haemorrhage (SAH) and delayed ischaemic dysfunction, to the clinical use of nimodipine, which has been largely documented among calcium inhibitors for its cerebrovascular properties.

This supplementum of Acta Neurochirurgica is based upon the papers presented. Some articles, however, have been extended to take account further results in order to present a broad view of the subject.

The Symposium started with two presentations concerning general aspects of SAH: in the first one, B. Voldby reviewed the pathophysiological events following SAH: if modifications of intracranial pressure (ICP) and decrease of cerebral blood flow (CBF) are the primary alterations, metabolic changes (particularly reduction in $CMRO_2$) contribute rapidly to disturbances of brain circulation. A variety of mechanisms may be responsible for the neurological dysfunction: most of them are directly related to the reduction in cerebral perfusion pressure (CPP) which may result in ischaemia. This fall in CPP (either due to elevation of ICP or reduction of mean arterial blood pressure) is observed under different pathophysiological conditions, discussed in the paper by A. D. Mendelow.

The second part of the Symposium was mainly related to the cerebro-vascular and direct cerebral effects of nimodipine (L. Brandt). By inhibiting calcium influx through calcium channels, calcium antagonists play a key role in the control of contractile activation of brain arteries. The selective cerebro-vascular effects can be demonstrated by the increase in CBF found in various experimental and clinical studies. In vasospasm, apart from this direct vasodilator action, a hypothesis has been proposed concerning a possible action on platelet aggregation, which however has not yet been proved. Besides the well-documented vascular effect, nimodipine may also exert a direct effect on nervous cells. The discovery of specific receptors in the human brain for dihydropyridine derivatives suggests such an action, confirmed by the blockage of calcium channel currents in endocrine cells.

The haemodynamic action of nimodipine can be evaluated by means of the transcranial doppler method. The report of Harders deals with clinical studies in human cerebral vasospasm following subarachnoid haemorrhage. Nimodipine does not seem to reduce the incidence of pathological blood flow velocities, indicative of vasoconstriction of the large vessels at the base of the brain. This is confirmed by several angiographic studies. However, critical and very high blood flow velocities were observed less frequently during nimodipine administration. The conclusion of the different transcranial doppler studies is that nimodipine reduces the severity of vasospasm but not its incidence and time-course. The question remains open, whether the reduction of flow velocity is caused by an increase in vessel diameter of the large arteries or whether it reflects a dilating effect on the small resistance vessels.

Most studies reported in this issue deal primarily with the particular effect of nimodipine on cerebral circulation. However, its systemic actions should not be disregarded. It is the purpose of the paper by A. Müller et al., to review the interactions between this drug and general anaesthesia. This study has been conducted during opiate anaesthesia for neurosurgical operations and confirms the vasodilating properties, i.e. the hypotensive action of nimodipine. In special circumstances, it may modify pulmonary circulation and gas exchange during anaesthesia. On the other hand, potentiation of analgesia as assumed in other studies has not been confirmed.

Therapeutic aspects constitute the third part of this volume. In a first chapter, R. H. Wilkins emphasizes, among various attempts to prevent vasospasm and to treat neurological deficit, three main directions: early operation to diminish the risk of rebleeding and to remove as much blood as possible from the basal cisterns, maintenance or elevation of circulating blood volume and maintenance or elevation of systemic blood pressure.

"Does nimodipine prevent ischaemic deficits after aneurysmal subarachnoid haemorrhage?" is the title of the review of recent clinical studies by J. M. Gilsbach. Trials with oral nimodipine, performed in a controlled and double-blind fashion, show obvious positive effects, but concern only a small series with delayed surgical treatment, which produce per se more unfavourable outcome and more ischaemic deficits than early surgery.

There is no randomized study with intravenous nimodipine and early surgery. However, several open and multicenter studies show that this treatment regimen is followed by a good outcome and a reduction of delayed ischaemic deficits to 1 to 7%. These results seem to be better than those obtained with oral treatment and delayed operation.

The influence of nimodipine in already established and symptomatic vasospasm is discussed in the last paper by F. Buchheit and P. Boyer. Mortality and severe morbidity were reduced in a randomized study conducted on 188 patients. This reduction is more pronounced when the treatment was started as early as possible; some open studies support these findings.

In conclusion, nimodipine is more effective in the prevention than in the treatment of established vasospasm. On the basis of actual knowledge and experience it seems that early surgery combined with prophylactic treatment represents the best option for patients after aneurysmal rupture.

J. Philippon
H. J. Reulen

Contents

Listed in Current Contents

Voldby, B., Pathophysiology of Subarachnoid Haemorrhage. Experimental and Clinical Data 1

Mendelow, A. D., Pathophysiology of Delayed Ischaemic Dysfunction After Subarachnoid Haemorrhage: Experimental and Clinical Data 7

Brandt, L., Andersson, K.-E., Ljunggren, B., Säveland, H., Ryman, T., Cerebrovascular and Cerebral Effects of Nimodipine—an Update 11

Harders, A., Gilsbach, J., Haemodynamic Effectiveness of Nimodipine on Spastic Brain Vessels After Subarachnoid Haemorrhage Evaluated by the Transcranial Doppler Method. A Review of Clinical Studies 21

Müller, H., Kafurke, H., Marck, P., Zierski, J., Hempelmann, G., Interactions Between Nimodipine and General Anaesthesia—Clinical Investigations in 124 Patients During Neurosurgical Operations 29

Wilkins, R. H., Attempts at Prevention and Treatment of Delayed Ischaemic Dysfunction in Patients with Subarachnoid Haemorrhage 36

Gilsbach, J. M., Nimodipine in the Prevention of Ischaemic Deficits After Aneurysmal Subarachnoid Haemorrhage. An Analysis of Recent Clinical Studies 41

Buchheit, F., Boyer, P., Review of Treatment of Symptomatic Cerebral Vasospasm with Nimodipine ... 51

Acta Neurochirurgica, Suppl. 45, 1–6 (1988)

Pathophysiology of Subarachnoid Haemorrhage

Experimental and Clinical Data

B. Voldby

Department of Neurosurgery, University Hospital, Aarhus, Denmark

Summary

When a saccular aneurysm suddenly ruptures the intracranial pressure (ICP) abruptly rises to reach a level at about the diastolic blood pressure in 1 to 2 minutes. Unless a haematoma is formed ICP will soon fall and reach a steady level in about 10 minutes. In the days following the initial SAH several pathophysiological events take place. Regional CBF and the cerebral metabolic rate of oxygen ($CMRO_2$) are reduced resulting in so-called luxury perfusion due to an uncoupling between flow and metabolism. The arteriovenous difference of oxygen is always reduced. $CMRO_2$ falls parallel to increasing severity of vasospasm. CBF below $20\,ml/100\,g/min$ in cases of severe diffuse spasm inevitably result in brain tissue infarction. The development of vasospasm, which reaches a maximum between the 5th and 9th day after SAH, is accompanied by CSF lactacidosis and intracranial hypertension. The reactivity of the cerebral arteries after SAH is often impaired. Cerebral autoregulation to arterial hypotension is disturbed even in mild cases, and globally fails in severe vasospasm. On the other hand the reactivity of the cerebral vasculature to changes in arterial PCO_2 is always preserved although reduced. Only in the presence of severe tissue acidosis will both modes of reactivity be damaged — so-called total vasoparalysis.

Keywords: Intracranial aneurysm; subarachnoid haemorrhage; intracranial pressure; regional cerebral blood flow; cerebral metabolic rate; autoregulation; vasospasm; vascular reactivity.

Introduction

Rupture of an intracranial saccular aneurysm is a dramatic and threatening event which may elicit a chain reaction eventually resulting in cerebral ischaemia. Most of the links in this chain are relatively well-known and well-understood, *e.g.* hydrocephalus, cerebral oedema, and haematoma, whereas cerebral arterial spasm or vasospasm is one of the most poorly understood phenomena in subarachnoid haemorrhage (SAH). Cerebral arterial spasm as revealed by angiography is now widely accepted as a major cause of delayed cerebral ischaemia that is at least as important as recurrent haemorrhage as a cause of mortality and morbidity after aneurysmal rupture[10]. The aetiology and pathogenesis of cerebral vasospasm is largely unknown. Attempts at preventing and treating vasospasm have been undertaken during the last 20 years without success[32]. A thorough knowledge of the pathophysiological mechanisms involved in the damaged cerebral circulation after aneurysmal rupture and in the development of vasospasm appears an absolute prerequisite for the rational therapy, in particular for the use of a new generation of vasoactive drugs, the calcium antagonists.

Many experimental studies and several clinical studies have been performed to elucidate the above mentioned problems. In the present survey some experimental data and recent clinical work pertinent to the subject of delayed cerebral dysfunction will be discussed.

Experimental Data

Experimental data concerning the effects of SAH on the cerebral haemodynamics and on the cerebral vasculature are numerous. Here, some studies relevant to

Table 1. *Subarachnoid Haemorrhage*
Pathophysiological Effects
Experimental Data

Early effects
Intracranial pressure rise
CBF reduction
CBV increase
Secondary hyperaemia
Energy-rich phosphates decrease
Segmental, localized vasospasm
Extracellular potassium accumulation
Extracellular calcium depletion

the clinical situation will be discussed. Since the pioneering work of Echlin[4] the vasoconstrictor effect of blood on the adventitial surface of the large cerebral arteries has been recognized. Vasospasm is a biphasic phenomenon consisting of an acute phase of maximally one hour and a chronic phase lasting from hours to days[2]. Simeone[19] confirmed this finding in a series of experimental SAH in the monkey. Prolonged vasospasm was produced by puncture of an intracranial artery or subarachnoid injection of autologous blood. Measurement of cerebral blood flow (CBF) revealed that only angiographic constriction of cerebral arteries to less than 50% of their control value was associated with significant reduction of CBF. This finding has later been confirmed in clinical studies[25, 29].

The most important effects of SAH in the acute experimental situation are shown in Table 1. The instantaneous injection of blood into the subarachnoid space, which resembles the rupture of an aneurysm in humans, makes the intracranial pressure (ICP) rise to the level of the arterial pressure followed by a return to a steady state value[20]. The pressure increase is quantitatively related to the amount of blood entering the system. CBF and cerebral metabolism decrease in the early hours after SAH[5]. This is preceeded by an initial increase in cerebral blood volume (CBV) as cerebral perfusion pressure (CPP) is reduced[7]. At this early stage segmental, localized spasm around the bleeding point may be seen. Of special interest for the use of calcium blocking agents in SAH is the study of the cerebral microcirculation showing that the application of blood to the cortex is followed by an extracellular potassium accumulation and calcium depletion[11].

The experimental findings of the delayed effects of SAH are also of interest to the clinical situation (Table 2). A protracted elevation of ICP results in a reduction of CPP. In addition, the development of diffuse vasospasm may reduce both CBF and cerebral metabolic rate of oxygen ($CMRO_2$)[8]. Cerebral autoregulation is globally depressed in animals of all clinical grades whereas CO_2 reactivity is depressed regionally corresponding to neurological deficit in poor-grade animals[12].

Clinical Data

Methodological Considerations

Our possibilities for measuring physiological parameters in patients with acute cerebrovascular disorders are limited by several factors. In the case of SAH from a ruptured intracranial aneurysm the risk of provoking a recurrent haemorrhage by the infliction of pain and discomfort to the patient must be considered. In Table 3 the parameters most often measured in clinical neurosurgery are shown. The continuous monitoring of the intraventricular pressure (IVP) is widely accepted for routine use. In the setting of SAH where some degree of hydrocephalus commonly develops initially the use or an indwelling ventricular catheter is preferable. This method also allows the concurrent measurement of CSF lactate and pH, and the performance of CSF drainage[23]. The risk of infection of the ventricular system should not be neglected, and a rate of infection of about 1% must be taken into account.

The quantitative measurement of regional CBF (rCBF) is obtained by the use of freely diffusible inert gases. The intracarotid 133-Xenon injection method has been used in several clinical series[9, 25] but the carotid puncture makes it unacceptable for repeated measurements in the same patient. The non-invasive 133-Xenon inhalation method has gained access to clinical work[30] but the calculation of the clearance curves is complicated by recirculation of isotope and artefacts from the

Table 2. *Subarachnoid Haemorrhage*
Pathophysiological Effects
Experimental Data

Delayed effects

Intracranial pressure increase
Cerebral vasospasm
CBF reduction
$CMRO_2$ reduction
Autoregulation impairment
CO_2-reactivity reduction
Central conduction time prolongation
CSF lactacidosis

Table 3. *Subarachnoid Haemorrhage*
Pathophysiological Effects
Clinical Parameters

Intracranial pressure (IVP)
CSF-Lactate/pH
rCBF Intra-arterial 133-Xe injection
     Inhalation 133-Xe
     SPECT
     PET
$AVDO_2$-$CMRO_2$
Cerebral autoregulation
CO_2-reactivity

---

upper airways. Recently, a single photon emission computed tomography method, which gives three-dimensional pictures of rCBF, has been used in SAH patients for repeated measurements in the evaluation of timing of surgery[14].

The clinical measurement of cerebral metabolism in SAH patients is difficult. CSF lactate gives a rather crude evaluation of cerebral ischaemia[23]. For the more accurate determination of cerebral metabolic rates of oxygen, glucose, or lactate the concurrent measurement of these metabolites in arterial and jugular vein blood is necessary[25]. An evaluation of the reactivity of the cerebral vasculature is possible but should be conducted with caution. Cerebral autoregulation is probably best tested by lowering the mean arterial blood pressure (MABP) by pharmacological methods (sodium nitroprusside) as the induction of systemic hypertension may provoke a recurrent haemorrhage. The reactivity of the cerebral arterioles to changes in arterial CO_2 tension may be tested by hyperventilation, either spontaneous or artificial[27].

Early Effects of Subarachnoid Haemorrhage

The primary event responsible for further pathophysiological changes is the rupture of the aneurysmal sac into the subarachnoid space. Measurements of this unique event have for natural reasons never been performed in humans. Nevertheless, we have several excellent recordings of IVP during recurrent haemorrhage from patients subjected to continuous monitoring of IVP. Conceivably this is similar to the initial rupture. In Fig. 1 such a pressure recording is shown. IVP abruptly rises to reach the distolic blood pressure and then in about 10 minutes falls to a steady state level the height of which seems to depend on the magnitude of the bleeding[16]. The severity of the bleeding is determined by several factors including intracranial coun-

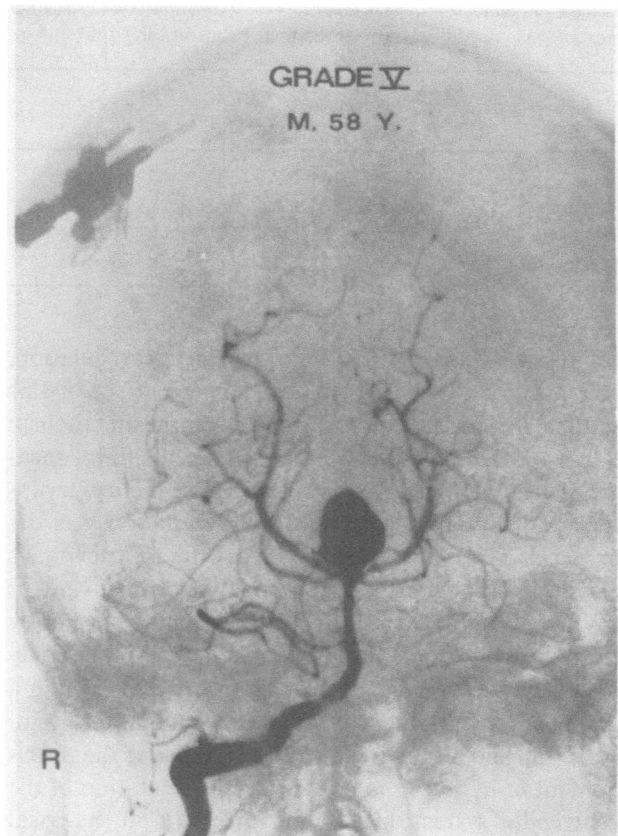

A

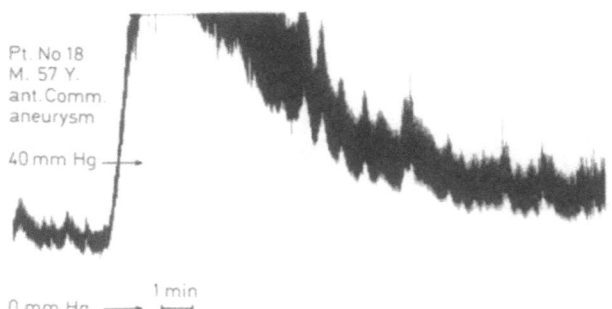

Fig. 1. Typical recording of the intraventricular pressure during recurrent haemorrhage. (From Voldby 1987[27])

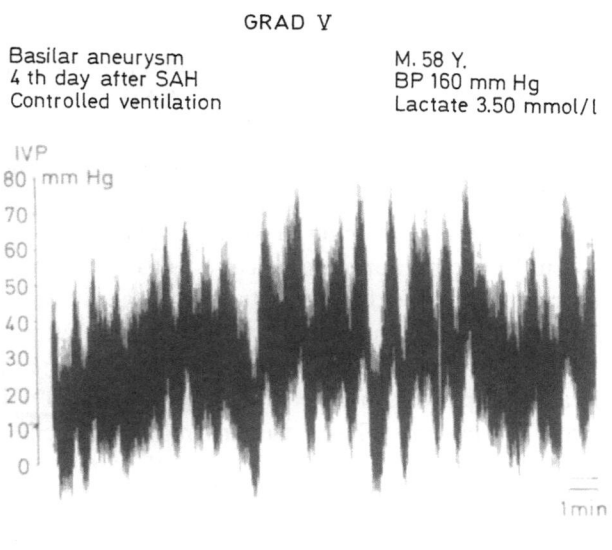

B

Fig. 2. (A) Angiography showing basilar artery aneurysm and diffuse vasospasm in 58-year-old male in clinical grade V. (B) Severely increased intraventricular pressure with large B-waves. CSF lactate elevated

Table 4. *Relation Between Cerebral Vasospasm and Cerebral Blood Flow, Cerebral Metabolism, and Intraventricular Pressure* (Values are mean ± SD. For abbreviations see text. From Voldby 1987[27])

Degree of vasospasm	n	Mean CBF ml/100 g/min	AVDO$_2$ ml %	CMRO$_2$ ml/100 g/min	IVP mmHg	CSF lactate mmol/l	CSF pH
None	11	44 ± 3.8	4.35 ± 1.02	2.11 ± 0.31	12 ± 5.3	1.92 ± 0.63	7.41 ± 0.05
Slight	12	44 ± 6.1	3.28 ± 1.27	1.57 ± 0.57	25 ± 13.0	2.90 ± 1.37	7.36 ± 0.04
Severe Focal	5	38 ± 6.4	4.30 ± 1.49	1.57 ± 0.37	31 ± 24.4	2.81 ± 0.41	7.35 ± 0.09
Severe Diffuse	10	21 ± 5.2	3.78 ± 1.58	0.82 ± 0.29	25 ± 9.2	3.19 ± 1.79	7.37 ± 0.04

terpressure, clot formation, cerebral autoregulation, and outflow of CSF. In particular, the effect of this sudden rise in ICP on CBF and cerebral metabolism is practically unknown. Nornes coincidentally measured arterial flow during intraoperative aneurysm rupture and found that cerebral autoregulation was defective[15]. The course of IVP during the first few days after the initial rupture has been studied thoroughly[21, 22]. In good-grade patients IVP is normal or slightly elevated while poor-grade patients have a moderately or severely elevated pressure. An example of IVP in a grade V patients is shown in Fig. 2. A pathophysiological mechanism of clinical importance is the development of hydrocephalus. This initial dilatation of the ventricular system due to clot formation in the basal cisterns and clogging of the arachnoid villi by shed blood corpuscles contributes to the raised ICP found in more than half of our patients[22].

Delayed Effects of Subarachnoid Haemorrhage

Although much remains to be elucidated regarding the production of vasospasm following SAH there is at hand enough data to hypothesize that blood, erythrocyte break down products or vasoactive compounds released from blood elements, brain tissue or the vascular wall may act singly or synergistically to induce spasm[10]. The pathophysiological effect of vasospasm on cerebral perfusion and metabolism was studied in 38 patients within the first 14 days after SAH (Table 4). Mean rCBF was normal or slightly reduced in grade II and III patients but decreased by at least 50% in grade IV patients. Slight vasospasm was associated with a normal or slightly reduced CBF while severe spasm was associated with severe reduction in rCBF (Fig. 3). Patients with severe diffuse spasm had rCBF values about 20 ml/100 g/min (Fig. 4). Regional changes were only found in patients with vasospasm. The angiographic demonstration of severe spasm correlated with the subsequent finding of infarcts in the territory of the spastic arteries on CT scanning or at autopsy.

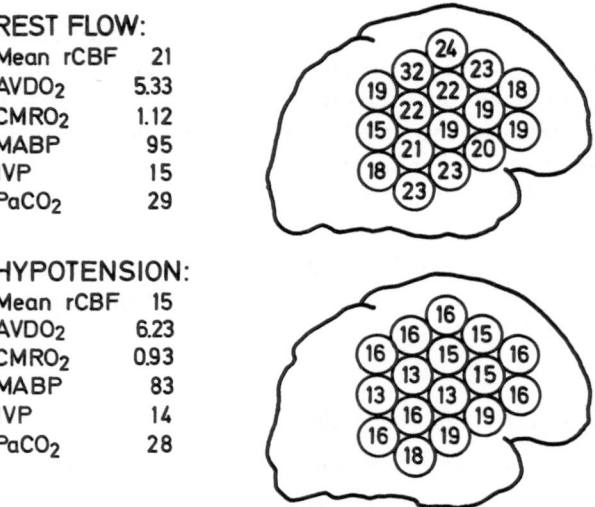

REST FLOW:
Mean rCBF	21
AVDO$_2$	5.33
CMRO$_2$	1.12
MABP	95
IVP	15
PaCO$_2$	29

HYPOTENSION:
Mean rCBF	15
AVDO$_2$	6.23
CMRO$_2$	0.93
MABP	83
IVP	14
PaCO$_2$	28

Fig. 3. 38-year-old male with right internal carotid aneurysm. CBF on day 8 showed global ischaemia. Autoregulation was impaired during short-lasting hypotension. Angiography revealed severe, diffuse vasospasm

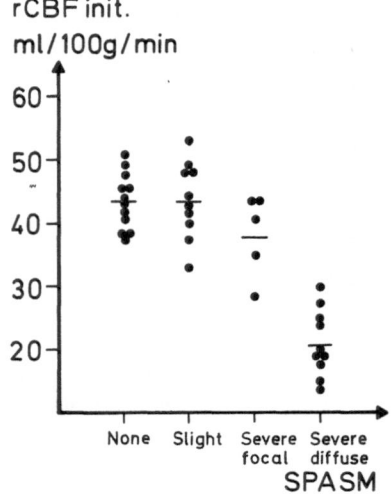

Fig. 4. Relation between rCBF and degree of vasospasm in 38 patients with ruptured intracranial aneurysm. (From Voldby *et al.* 1985[25])

Arteriovenous difference of oxygen was measured during resting flow. $AVDO_2$ was equally reduced in all patients irrespective of the degree of vasospasm, whereas $CMRO_2$ reductions were closely related to the degree of vasospasm. $CMRO_2$ was relatively more reduced than rCBF indicating an uncoupling between flow and metabolism. This uncoupling was more pronounced in patients in good clinical condition while CBF was more closely related to the decreased metabolic demands of the ischaemic brain tissue in grade IV and V patients. Cerebral autoregulation was in our study tested in 25 patients by reducing MABP shortly by 10 to 20%[27]. Patients without vasospasm had a normal autoregulation. Patients with slight vasospasm often had focal impairment, and patients with severe vasospasm always had impaired autoregulation, focally or globally depending on the extent of the spasm. Increased IVP and CSF lactacidosis were often associated with impaired autoregulation. The ability of the cerebral vasculature to react to changes on $PaCO_2$ was tested in 29 patients by hyperventilation. In all patients hyperventilation was followed by a reduction in mean rCBF of between 1 and 4% per mm of Hg of $PaCO_2$ reduction. However, only reductions of less than 2% were associated with cerebral ischaemia. Thus all patients with severe spasm had a globally depressed CO_2 response[27].

Discussion

When an intracranial saccular aneurysm suddenly bursts blood is forced under a pressure of approximately 100 mmHg into the subarachnoid space where a pressure of between 0 and 10 mmHg prevails under normal circumstances. As far as we know from human physiology similar shortlasting rises in ICP, *e.g.* under straining, are harmless to the cerebral circulation. This is probably due to the cerebral autoregulation which keeps flow constant within certain limits when CPP falls. Why is it then that SAH is not tolerated and is so often followed by clinical deterioration?

The most obvious reason is the presence of blood in the subarachnoid space. Studies of the cerebral cortical microcirculation after the application of blood have shown that capillary vasospasm leads to marked tissue hypoxica. This decrease in perfusion pressure of the capillary bed could not be compensated for by increasing the arterial systemic blood pressure[31]. Hence, the initial aneurysmal rupture may imply at least three significant pathophysiological mechanisms: 1. Reduction in CPP. 2. Widespread cortical microcirculatory changes. 3. Cerebral arterial and arteriolar

constriction. The last point has not been demonstrated in humans but from experimental studies we know that angiographically demonstrable vasospasm develops within 5 minutes of the introduction of blood into the basal cisterns[2]. It seems possible that these factors are responsible for the production of cerebral ischaemia in the acute stage. From a therapeutic point of view, however, the clinical course over the days following the initial haemorrhage are more important. IVP is elevated in patients in poor clinical condition and in patients with vasospasm. Attempts at improving their CCP by drainage of CSF have been successful[21] but also implies a certain risk if a repeat haemorrhage occurs during drainage[24].

Delayed cerebral ischaemia in SAH is generally used almost synonymous with delayed cerebral vasospasm or symptomatic vasospasm indicating a clinical state characterized by a deterioration of neurological function associated with angiographic narrowing of more than 50% of the normal diameter of the cerebral arteries appropriate to the neurological deficit. The quantitation of vasospasm is significant. Slight vasospasm (a reduction in arterial diameter of between 25 and 50%) is not associated with clinically manifest cerebral ischaemia[25]. In this connection, however, it is important to stress that slight vasospasm is not a totally innocent phenomenon. A progression to severe spasm is possible, and distinct abnormalities in focal cerebral autoregulation have been found in patients with slight spasm but in satisfactory clinical condition[26].

Severe vasospasm as defined above is on the other hand a serious condition which almost inevitably leads ischaemia and cerebral infarction[25]. Thresholds of cerebral ischaemia have been defined lately. Under experimental conditions CBF below 23 ml/100 g/min results in reversible neurological deficit which deteriorates to complete deficit at flow values below 10 ml/100 g/min[13]. Brain tissue with flows ranging between 10 and 15 ml/100 g/min may still be viable – the so-called ischaemic penumbra[1]. Reversible ischaemia due to vasospasm has been demonstrated by the positron emission tomography method[18].

Main topics of the present symposium are prevention and treatment of delayed ischaemic dysfunction. Several clinical studies point to cerebral vasospasm as a major cause of delayed ischaemia[10] but other factors should not be overlooked in the pre- and postoperative period[17]. The prevention of ischaemic dysfunction is the most significant goal to be achieved. The amount and distribution of blood on the initial CT scanning may be helpful in identifying the patients at risk of developing vasospasm. As regards the treatment of manifest neurological dysfunction pathophysiological considerations are of the utmost importance. There is ample experimental evidence that nimodipine effectively dilates constricted cerebral arteries *in vivo*[26]. In cerebral ischaemia the effect of vasodilating agents may be unpredictable. A risk of only dilating the vessels around the ischaemic area may result in intracerebral steal. This inexpedient mode of action of a calcium antagonist has been found in acute stroke patients[28]. The monitoring of pathophysiological parameters in patients with delayed ischaemic dysfunction may avoid unwanted effects. The finding of manifest cerebral infarction, a defect of autoregulation, intracranial hypertension and severe CSF lactacidosis should warrant caution in the use of vasodilators. However, in a clinical series of 54 patients with different cerebrovascular disorders including SAH there was no evidence of a steal phenomenon[6].

References

1. Astrup J (1982) Energy-requiring cell functions in the ischaemic brain. J Neurosurg 56: 484–497
2. du Boulay G, Symon L, Ackerman RH *et al* (1973) The reactivity of the spastic arteries. Neuroradiology 5: 37–39
3. Brawley BW, Strandness Jr DE, Kelly WA (1968) The biphasic response of cerebral vasospasm in experimental subarachnoid haemorrhage. J Neurosurg 28: 1–8
4. Echlin F (1971) Experimental vasospasm, acute and chronic due to blood in the subarachnoid space. J Neurosurg 35: 646–656
5. Fein JM (1975) Cerebral energy metabolism after subarachnoid haemorrhage. Stroke 6: 1–8
6. Gaab MR, Haubitz L, Brawanksi A *et al* (1985) Acute effects of nimodipine on the cerebral blood flow and intracranial pressure. Neurochirurgia 28: 93–99
7. Grubb RL Jr, Raichle ME, Ratcheson RA (1975) Effects of increased intracranial pressure on cerebral blood volume, blood flow and oxygen utilization in monkeys. J Neurosurg 45: 385–398
8. Hashi K, Meyere JS, Shinmura S *et al* (1972) Cerebral haemodynamic and metabolic changes after experimental subarachnoid haemorrhage. J Neurol Sci 17: 1–14
9. Heilbrun MP, Olesen J, Lassen NA (1972) Regional cerebral bloow flow studies in subarachnoid haemorrhage. J Neurosurg 37: 36–44
10. Heros RC, Zervas NT, Varsos V (1983) Cerebral vasospasm after subarachnoid haemorrhage. An update. Ann Neurol 14: 599–608
11. Hubschman OR, Kornhauser D (1982) Effect of subarachnoid haemorrhage on the extracellular microenvironment. J Neurosurg 56: 216–221
12. Jakubowski J, Bell BA, Symon L *et al* (1982) A primate model of subarachnoid haemorrhage: Changes in regional cerebral blood flow, autoregulation, carbondioxide reactivity, and central conduction time. Stroke 12: 601–611
13. Jones TH, Morawetz RB, Crowell RM *et al* (1982) Thresholds of focal cerebral ischaemia in awake monkeys. J Neurosurg 54: 773–782
14. Mickey B, Vorstrup S, Voldby B *et al* (1984) Serial measurements of regional cerebral blood flow in patients with subarachnoid haemorrhage using 133-Xenon inhalation and emission computed tomography. J Neurosurg 60: 916–922
15. Nornes H (1978) Cerebral arterial flow dynamics during aneurysm haemorrhage. Acta Neurochir (Wien) 41: 39–48
16. Nornes H, Magnæs B (1972) Intracranial pressure in patients with ruptured saccular aneurysm. J Neurosurg 36: 537–547
17. Peerless SJ (1979) Pre- and postoperative management of cerebral aneurysms. Clin Neurosurg 26: 209–231
18. Powers WJ, Grubb RL, Baker RP *et al* (1985) Regional cerebral blood flow and metabolism in reversible ischaemia due to vasospasm. J Neurosurg 62: 539–546
19. Simeone FA, Trepper PJ, Brown DJ (1972) Cerebral blood flow evaluation of prolonged experimental vasospasm. J Neurosurg 37: 302–311
20. Steiner L, Löfgren J, Zwetnow NM (1975) Characteristics and limits of tolerance in repeated subarachnoid haemorrhage in dogs. Acta Neurol Scand 52: 241–267
21. Sundbärg G, Ponten U (1976) ICP and CSF absorption impairment after subarachnoid haemorrhage. In: Beks JWF, Bosch DA, Brock M (eds) Intracranial pressure III. Springer, Berlin Heidelberg New York, pp 139–146
22. Voldby B, Enevoldsen EM (1982) Intracranial pressure changes following aneurysm rupture. Part 1: Clinical and angiographic correlations. J Neurosurg 56: 186–196
23. Voldby B, Enevoldsen EM (1982) Intracranial pressure changes following aneurysm rupture. Part 2: Associated cerebrospinal fluid lactacidosis. J Neurosurg 56: 197–204
24. Voldby B, Enevoldsen EM (1982) Intracranial pressure changes following aneurysm rupture. Part 3: Recurrent haemorrhage. J Neurosurg 56: 784–789
25. Voldby B, Enevoldsen EM, Jensen FT (1985) Regional cerebral blood flow, intraventricular pressure, and cerebral metabolism in patients with ruptured intracranial aneurysm. J Neurosurg 62: 48–58
26. Voldby B, Petersen OF, Buhl M *et al* (1984) Reversal of cerebral arterial spasm by intrathecal administration of a calcium antagonist (nimodipine). Acta Neurochir (Wien) 70: 243–254
27. Voldby B (1987) Alterations in vasomotor reactivity in subarachnoid haemorrhage. In: Wood JH (ed) Cerebral blood flow. Physiologic and clinical aspects. McGraw-Hill, New York, pp 402–412
28. Vorstrup S, Andersen A, Blegvad N *et al* (1986) Calcium antagonist (PY 108-068) treatment may further decrease flow in ischaemic areas in acute stroke. J Cereb Blood Flow Metab 6: 222–229
29. Weir B, Grace M, Hansen J *et al* (1978) Time course of vasospasm in man. J Neurosurg 48: 173–178
30. Weir B, Menon D, Owerton T *et al* (1978) Regional cerebral blood flow in patients with aneurysms: Estimation by Xenon 133 inhalation. Can J Neurol Sci 5: 301–305
31. Wiernsperger N, Schultz U, Gygax P (1981) Physiological and morphometric analysis of the microcirculation of the cerebral cortex under acute vasospasm. Stroke 12: 624–627
32. Wilkins RH (1987) Cerebral vasospasm. Prevention and treatment. In: Wood JH (ed) Cerebral blood flow. Physiologic and clinical aspects. McGraw-Hill, New York, pp 603–612

Correspondence: Bo Voldby, M.D., Ph.D., Department of Neurosurgery, Aarhus Kommunehospital, DK-8000 Aarhus C, Denmark.

Acta Neurochirurgica, Suppl. 45, 7–10 (1988)

Pathophysiology of Delayed Ischaemic Dysfunction After Subarachnoid Haemorrhage: Experimental and Clinical Data

A. D. Mendelow

Department of Neurosurgery, University and General Hospital, Newcastle Upon Tyne, U.K.

Summary

Cerebral arterial spasm from subarachnoid haemorrhage (SAH) may be associated with the clinical syndrome of delayed cerebral ischaemic dysfunction, but the two conditions are by no means synonymous. Patients in good clinical condition may be seen with severe vasospasm and vice versa. A variety of mechanisms may be responsible for the neurological dysfunction: current evidence indicates that most of these mechanisms produce a reduction in cerebral blood flow (CBF). In the early stages after SAH, there is a loss of autoregulation so that reduction in cerebral perfusion pressure (CPP) may produce ischaemia. This fall in CPP may be due to elevated intracranial pressure (ICP) or reduced mean arterial blood pressure (MABP). Delayed cerebral ischaemic dysfunction following SAH is a clinical syndrome which may also be caused by re-bleeding, hydrocephalus, dehydration, reduced cardiac output and/or blood pressure, hyperglycemia or epilepsy.

Experimental evidence has indicated that at the time of an intracranial haemorrhage there is a profound and extensive focal ischaemic insult. The severity of the ischaemia depends upon the nature, size and rapidity of onset of haemorrhage. The focal ischaemic lesion around an intracerebral haemorrhage becomes smaller with the passage of time and is the result of two interacting phenomena:

a) The physical properties of the haemorrhage result in increased local tissue pressure around the lesion. This produces squeezing of the microcirculation and focal ischaemia.

b) Vasoconstrictor elements in blood produce spasm which may further reduce CBF, sometimes even remotely from the lesion.

Treatment of delayed ischaemia therefore depends upon its cause. If the clinical problem is reduced CPP, then restoration of CPP (with hypervolemia or hypertension) improves CBF provided that the ischaemic lesion has not proceeded to the stage of blood brain barrier (BBB) damage with brain oedema. In such a late situation, increasing CPP may aggravate the picture. The particular clinical therapy must therefore be appropriate and directed at the particular stage of evolution of the ischaemic lesion.

Keywords: Subarachnoid haemorrhage; ischaemia; vasospasm; delayed ischaemic dysfunction; pathophysiology.

Introduction

Delayed cerebral ischaemic dysfunction is a clinical syndrome which occurs in patients following subarachnoid haemorrhage. It is a multifactorial disorder for which there is no single treatment or cure. This clinical syndrome should not be confused with the angiographic appearance of narrow arteries which are frequently constricted because of smooth muscle contraction (vasospasm). Vasospasm may produce ischaemia, but this is not always the case: patients in good clinical condition may have severe vasospasm and vice versa. The neurological dysfunction is more likely to be the result of reduced cerebral blood flow (CBF). In the early stages after subarachnoid haemorrhage (SAH) there is loss of autoregulation so that any reduction in the cerebral perfusion pressure (CPP) may result in ischaemia. The clinical syndrome may be due to a variety of pathophysiological factors (Table 1).

Successful treatment depends on recognition of which factor is responsible for the clinical problem, and its correction.

Causes of Delayed Cerebral Ischaemic Dysfunction

1. Vascular Narrowing

This may be mechanical, the result of smooth muscle constriction (vasospasm) or neurogenic.

Table 1. *Delayed Cerebral Ischaemia*

Pathophysiological factors
1 Vascular narrowing
2 Reduced CBF
3 Rebleeding
4 Hydrocephalus
5 Dehydration
6 Reduced cardiac output and/or BP
7 Hyperglycaemia
8 Epilepsy

Mechanical constriction may occur with brain displacement or herniation. For example, tentorial herniation may cause compression of the posterior cerebral artery. Alternatively organizing thrombus may prevent dilatation of the cerebral arteries after a period of initial vasoconstriction at the time of haemorrhage. Ultrastructural changes in the vessel wall have been reported by Alskne[1]. These include loss of the endothelium, subendothelial proliferation, disruption of the internal elastic lamina and smooth muscle necrosis. He prefers the term vasonecrosis to vasospasm, but each probably represents a different cause for vascular narrowing.

Vascular smooth muscle constriction (*vasospasm*) is seen in about a third of patients after subarachnoid haemorrhage and may be due to a wide variety of different vasoconstrictor substances released from blood at the time of the haemorrhage. These include various catecholamines, indolalkylamines, haem and its derivatives, prostaglandins, thromboxanes and arachidonic acid as well as various kinins and other polypeptides. The relative importance of these various "spasmogens" is uncertain. The wide range of drugs which are advocated for the treatment of vasospasm[17], and the fact that no single agent has proved universally successful, suggests that perhaps there is no single "spasmogen" to blame in SAH.

Vascular narrowing may also be the results of *neurogenic* smooth muscle constriction due to release of neurotransmitters from adrenergic[10] or cholinergic[4] nerves on cerebral vessels. It is likely that remote vasospasm (in other parts of the cerebral arterial tree or indeed in the coronary circulation) may be mediated by these neurogenic mechanisms. It has been shown that SAH results in degeneration of cerebrovascular adrenergic fibres[12], and this may produce denervation hypersensitivity so that a second haemorrhage results in increased sensitivity of cerebrovascular smooth muscle to vasoconstriction. By contrast, Delgardo *et al.*[3] have shown that denervation of the central catecholamine neurons with 6-hydroxydopamine in rats prevents delayed or chronic vasospasm. The controversy about the role of innervation in the pathogenesis of vasospasm therefore continues. There is however no doubt that neurogenic vasoconstriction is a cause of vascular narrowing in some patients.

Provided that there is an adequate perfusion pressure, and provided that the vasospasm is not too severe, there may be no effects on CBF. For vascular narrowing to produce a clinical effect, there must be an associated reduction in CBF as described below.

2. Reduced Cerebral Blood Flow (CBF)

Normally autoregulation maintains CBF in the face of changing blood pressure. However, loss of autoregulation occurs in patients following subarachnoid haemorrhage. This has been demonstrated preoperatively[13], intraoperatively[5] and postoperatively[6]. In all these situations, CBF becomes passively dependent upon the CPP. Any reduction in MABP or rise in the ICP therefore produces a change in CBF. While changes in CPP may not normally affect CBF, when there is vascular narrowing and loss of autoregulation, small changes in CPP may result in changes in CBF of sufficient magnitude to produce the clinical syndrome of delayed ischaemic dysfunction. Reduction in the cerebral metabolic rate may be associated with reduced CBF. Two factors may produce reduced CBF (*i.e.,* other than a fall in CPP):

a) Dynamic Vasospasm

Changes in arterial calibre as a result of altering smooth muscle tone may occur and are most likely to be associated with episodes of recurrent haemorrhage. Such changes (dynamic vasospasm) may be seen on subsequent angiograms, although these are now seldom performed.

b) Increased Tissue Pressure

Increased tissue pressure may occur in those patients who have large volumes of blood, particularly where there is intracerebral haemorrhage related to a ruptured aneurysm. Experimental studies in rats[7] have shown that a small intracerebral haematoma produces a pro-

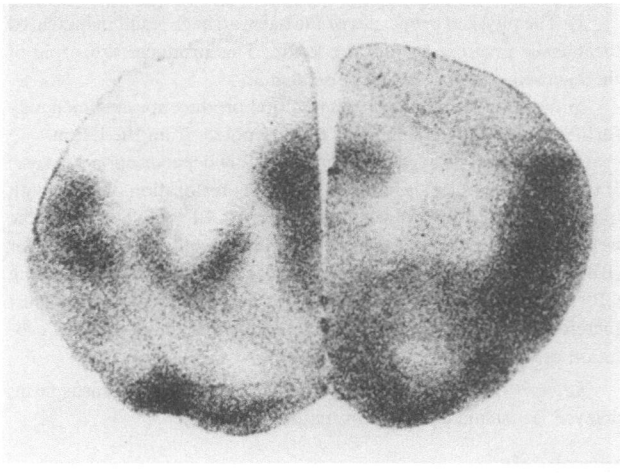

Fig. 1. Autoradiograph showing the large ischaemic area that surrounds a small haemorrhage in the caudate nucleus

found and extensive focal ischaemic insult (Fig. 1). The severity of the ischaemia depends upon the nature, size and rapidity of onset of haemorrhage. The focal ischaemic lesion becomes smaller with the passage of time and is the result of two interacting phenomena:

a. The physical properties of the haemorrhage result in increased local tissue pressure with squeezing of the microcirculation and focal ischaemia. This has now been demonstrated in a purely mechanical model of intracerebral haemorrhage in the rat, where a microballoon is inflated in the caudate nucleus[15]. This model eliminates the vasoconstrictor elements in blood and the ischaemic lesion is smaller than is seen with the injection of an equivalent amount of blood[8].

b. Vasoconstrictor elements in blood therefore produce additional vasoconstriction which may further reduce CBF sometimes even remotely from the lesion[9].

The reduction in CBF around the haematoma produces a central zone of infarction surrounded by an ischaemic penumbra. There is a rapid accumulation of calcium ions in infarcted cerebral tissue[14] and calcium channel blockers may prevent this accumulation and reduce infarct size. With experimental intracerebral haemorrhage there is a reduction in the volume of cerebral infarct surrounding the haematoma when animals are pretreated with the calcium antagonist nimodipine[16]. This reduction in infarct size was significant in the balloon inflation model but not with intracerebral haemorrhage, probably due to the greater variability in the haemorrhage model. The fact that a significant reduction in infarct size is shown with the mechanical balloon model indicates that calcium channel blockers may influence infarct size by preventing the rapid accumulation of intracellular calcium that occurs with ischaemia, rather than by relieving vasospasm. In the balloon model the vasospastic elements were not present, and therefore the smooth muscle relaxant properties of calcium channel blockers are unlikely to be exerting their effect by reducing vasospasm.

3. Rebleeding

A re-bleed from an aneurysm may be confused with delayed cerebral ischaemia, but can easily be differentiated from it by repeating the CT scan. The re-bleed may cause an acute rise in ICP which may reduce CPP to almost zero[11]. If a re-bleed is associated with an intracerebral haematoma, the increased tissue pressure around the haematoma may produce a reduction in CBF (see above). A re-bleed may also release further spasmogens, and if this follows denervation, then de-

nervation hypersensitivity may be responsible for further vascular narrowing.

4. Hydrocephalus

The increased ICP associated with hydrocephalus may lead to a fall in CPP with a further fall in CBF. Hydrocephalus can be easily excluded on a CT scan.

5. Dehydration

There are several explanations for dehydration in patients with ruptured intracranial aneurysms. The result will be a reduced circulating blood volume and increased blood viscosity. Both of these will further tend to reduce CBF. The causes of dehydration are a poor intake, the use of osmotic and other diuretics, the use of vasopressor agents which increase renal plasma flow, also producing a diuresis, and, occasionally, diabetes insipidus.

6. Reduced Cardiac Output and/or Blood Pressure

Almost any arrhythmia can occur following subarachnoid haemorrhage. Arrhythmias may result in reduced MABP with a fall in CPP. Recently Andreoli et al.[2] reported that 91% of patients studied prospectively within 48 hours of a subarachnoid haemorrhage will have some form of arrhythmia. Myocardial damage is often subendocardial but may be more extensive and this too may reduce cardiac output. Dehydration, the use of anti hypertensives, and other drugs may also lead to hypotension. Particular care should be exercised when calcium antagonist agents are used because of their potential hypotensive side effects. Intraoperative hypotension may prove dangerous in patients after a

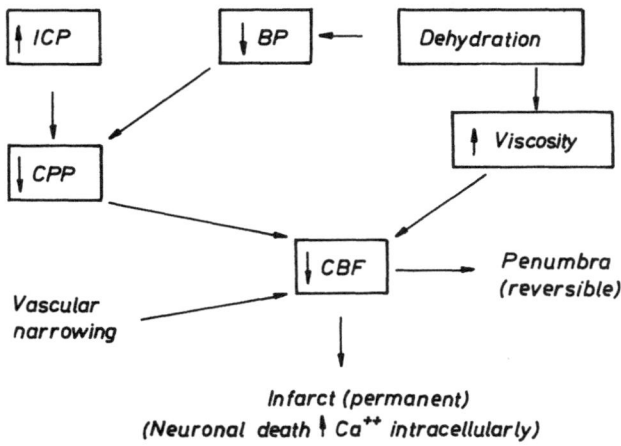

Fig. 2. Pathophysiological events that follow SAH and lead to delayed cerebral ischaemia

recent haemorrhage. Consequently with earlier surgery, clinicians have abandoned intraoperative hypotension, preferring temporary clips if necessary.

Finally, hypoglycemia and epilepsy may contribute to clinical deterioration in patients with subarachnoid haemorrhage and these complications should be considered in the differential diagnosis of delayed ischaemic dysfunction.

Conclusions

While vascular narrowing is often associated with delayed cerebral ischaemia, the final common pathway by which a permanent infarct or an ischaemic penumbra occurs following a subarachnoid haemorrhage is by a reduction of CBF. (The interplay of the various factors is summarized in Fig. 2.) The clinical syndrome of delayed cerebral ischaemic dysfunction should not be called "vasospasm". Avoidance of the latter term should encourage clinicians to consider the wider causes of clinical neurological deterioration.

References

1. Alksne JF, Branson PJ (1979) Prevention of experimental subarachnoid haemorrhage induced intracranial arterial vasonecrosis with phosphodiesterase inhibitor phthalazinol (EG-626). Stroke 10: 638–644
2. Andreoli A, Pasquale GD, Pinelli G, Grazi P, Tognetti F, Testa C (1987) Subarachnoid haemorrhage: frequency and severity of cardiac arrhythmias. Stroke 18: 558–564
3. Delgado TG, Arbab MA-R, Rosengrene E, Svendgaard NA (1987) The effect of neo-natal 6 hydroxy dopamine treatment on experimental vasospasm following a subarachnoid haemorrhage in the rat. J_CBF Metabol 7: 289–294
4. Denn MJ, Stone HL (1976) Cholinergic innervation of monkey cerebral vessels. Brain Res 113: 394
5. Farrar JK, Gamache FW, Ferguson GG, Drake CG (1981) Cerebral blood flow (CBF) in profound intraoperative hypotension: correlation of pre- and postoperative measurements. J CBF Metabol 1: 520–521
6. Mendelow AD, Dharker S, Patterson J, Nath F, Teasdale GM (1986) The dopamine withdrawal test following surgery for intracranial aneurysms. J Neurol Neurosurg Psychiatry 49: 35–38
7. Mendelow AD, Bullock R, Teasdale GM, Graham DI, McCulloch J (1984) Intracranial haemorrhage induced at arterial pressure in the rat: Part 2. Short term changes in local cerebral blood flow measured by autoradiography. Neurol Res 6: 189–193
8. Mendelow AD, Bullock R, Nath FP, Jenkins A, Kingman T, Teasdale GM (1986) In: Miller JD, Teasdale GM, Rowan JO, Galbraith SL, Mendelow AD (eds) Intracranial pressure VI. Springer, Berlin Heidelberg New York, pp 515–520
9. Nath FP, Jenkins A, Mendelow AD, Graham DI, Teasdale GM (1986) Early haemodynamic changes in experimental intracerebral haemorrhage. J Neurosurg 65: 697–703
10. Neilsen KC, Owman Ch (1967) Adrenergic innervation of pial arteries related to the circle of Willis in the cat. Brain Res 773
11. Nornes H (1975) Monitoring of patients with intracranial aneurysms. Clin Neurosurg 21: 321–331
12. Peerless SJ, Yaşargil MG (1971) Adrenergic innervation of the cerebral blood vessels in the rabbit. J Neurosurg 35: 148
13. Pickard JD, Read DH, Lovick AHJ (1986) Preoperative assessment of cerebrovascular reactivity following subarachnoid haemorrhage – clinical correlations. In: Auer L (ed) Cerebral aneurysm surgery in the acute stage. Springer, New York
14. Rappaport ZH, Young W, Flamm ES (1987) Regional brain calcium changes in the rat: middle cerebral artery occlusion model of ischaemia. Stroke 18: 760–764
15. Sinar EJ, Mendelow AD, Graham DI, Teasdale GM (1987) Experimental intracerebral haemorrhage: effects of a temporary mass lesion. J Neurosurg 66: 568–576
16. Sinar EJ, Mendelow AD, Graham DI, Teasdale GM (1987) Intracerebral haemorrhage: the effect of pretreatment with nimodipine on the volume of ischaemic damage. J CBF Metabol 7 [Suppl 1]: 158
17. Wilkins RH (1980) Attempt at prevention or treatment of intracranial arterial spasm: a survey. Neurosurgery 6: 198–210

Correspondence: Dr. A. D. Mendelow, Department of Neurosurgery, University and General Hospital, Newcastle Upon Tyne, U.K.

Acta Neurochirurgica, Suppl. 45, 11–20 (1988)

Cerebrovascular and Cerebral Effects of Nimodipine — an Update

L. Brandt, K.-E. Andersson[1], B. Ljunggren, H. Säveland, and **T. Ryman[1]**

Departments of Neurosurgery and [1]Clinical Pharmacology, University Hospital, Lund, Sweden

Summary

A survey is given on the vascular and neuronal effects of calcium antagonists under physiological and pathological conditions. Special emphasis is put on vasospasm caused by subarachnoid haemorrhage and on postischaemic cerebral hypoperfusion following different forms of cerebral ischaemia, and on the attempts to influence these phenomena pharmacologically.

Regarding its neuronal effects it seems likely that nimodipine potently blocks calcium entry during pathological conditions like cerebral ischaemia and spreading cortical depression. Positive effects also have been seen during epileptic seizures and withdrawal syndromes, whereas calcium entry under normal physiological conditions does not appear to be affected. Possible clinical consequences are discussed.

Keywords: Calcium antagonists; Nimodipine; vasospasm; ischaemia; epilepsy; withdrawal syndrome.

Introduction

Calcium is known to be an important link between electrical or chemical stimulation and physiological responses in a great variety of contractile and secretory cells. The extracellular calcium concentration is roughly about 10,000 times higher than that inside the cell. This gradient is achieved by the limited permeability of the resting plasma membrane to calcium, as well as by cellular mechanisms for active calcium extrusion. The uptake and storage of calcium by intracellular organelles contribute to maintain a low ($< 10^{-7}$) intracellular free calcium concentration. Calcium entry from the extracellular medium can occur through voltage- or receptor-operated calcium channels[11]. Voltage-dependent calcium channels are the best characterized and are widely distributed and present in all types of muscle endocrine secretory cells, neurones and glia[33]. Calcium channels play a key role for the availability of calcium for contraction in smooth muscle cells and for triggering neurotransmitter release in nerve terminals and controlling secretion of hormones from endocrine cells.

Multiple calcium channel subtypes have been identified in neurones, skeletal, smooth and cardiac muscle as well as in pituitary cells. The basis for distinguishing among channel subtypes includes differences in the membrane potential at which channels are activated, in their tendency to inactivate and in their pharmacological sensitivity (see *e.g.*, ref.[31]). In neurones of the dorsal root ganglion, three types of calcium channel with different sensitivity to calcium antagonists (see below) have been described[73]. The existence of multiple calcium channel subtypes may be related to the diversity of functions that are mediated in contractile, neural and secretory cells.

Calcium Antagonists

Drugs which can affect the movement of calcium, may be subdivided into two groups: 1. inhibitors of calcium movement, and binding, and 2. facilitators of calcium movement[97]. Substances which block calcium channels, "calcium antagonists" belong to the first group and are the only ones which hitherto are of therapeutic importance.

The WHO committee suggested a classification of calcium antagonists based on selectivity for slow calcium channels[97]: those selective for slow channels (verapamil and derivatives, dihydropyridines, diltiazem) and those non-selective for slow channels (difenylpiperazines, *e.g.*, flunarizine, prenylamine derivatives and others). Nimodipine is a calcium antagonist of the dihydropyridine type (Fig. 1).

Cerebrovascular Effects of Calcium Antagonists

Calcium antagonists are generally more potent inhibitors of calcium influx through voltage-dependent than receptor-operated channels. They are also more potent in smooth and cardiac muscle than in skeletal muscle.

In the circulation, they preferentially affect coronary and cerebral vascular smooth muscle cells[5]. This has been documented in several studies both *in vitro*[2, 10, 13, 14, 39, 46, 67, 93, 94, 104] and *in vivo*[7, 43, 48].

The mechanisms behind a selective cerebrovascular effect of calcium antagonists have not been established, but may be related to a greater dependence on extracellular calcium for contractile activation of cerebral rather than of peripheral vessels. McCalden and Bevan[57] concluded that K^+ induced contraction in isolated rabbit basilar arteries by using a single calcium pool, probably of extracellular origin. Noradrenaline (NA) and serotonin (5-HT) also primarily utilized extracellular calcium. Skärby *et al.*[86] found that in cat middle cerebral artery, both K^+- and noradrenaline-induced contractions were almost exclusively dependent on the presence of calcium in the extracellular medium, and that activation occurred through pathways sensitive to calcium antagonists. This is in some contrast with findings in isolated human pial vessels[13] where treatment in a calcium-free medium for 30 minutes and exposure to nifedipine or nimodipine markedly reduced K^+, but to a lesser extent NA- and 5-HT-induced contractions. It was suggested that this was due to the amines also using intracellulary stored calcium for their contraction. Uski and Andersson[95] found that in calcium-free medium $PGF_{2\alpha}$ induced a biphasic contraction in feline basilar arteries, probably by releasing cellularly bound calcium from two different stores. Prostaglandin-induced release of intracellular calcium in cerebral arteries was suggested also by Sasaki *et al.*[81]. Brandt *et al.*[13] found that in isolated human pial arteries, concentrations of nimodipine which abolished K^+-induced contractions, reduced $PGF_{2\alpha}$-induced contractions by only about 60%. Nosko *et al.*[71] studied isolated cerebral arteries from monkey, dog and man. In all species tested, prostaglandin $F_{2\alpha}$ was poorly antagonized also by high concentrations of nimodipine. Available information suggests that the requirement of extracellular vs intracellular calcium for contractile activation of brain arteries is dependent on what agent is used for the activation.

The effects of nimodipine on cerebral blood flow was studied by Haws and coworkers[38] in anaesthetized and unanaesthetized rabbits using the microsphere method. At low doses which have little effect on systemic blood pressure, nimodipine caused a selective increase in cerebral and myocardial blood flow. Nimodipine increased blood flow in all regions of the brain. There was no change in cerebral O_2 consumption and the increased blood flow was interpreted as the result of a direct vasodilator effect and not secondary to increased cerebral metabolism.

In a series of experiments in lightly restrained, conscious rats, Mohammed *et al.*[65] studied the effects of a continuous i.v. infusion of nimodipine on local CBF and local cerebral glucose utilization using the quantitative autoradiographic techniques. The nimodipine infusion produced only a small reduction in mean arterial blood pressure, and did not alter the rate of glucose utilization in any of the regions examined. By contrast, in 24 regions, CBF was increased significantly by 39–84% from control levels.

A selective action on cerebral vessels may contribute to a beneficial effect of calcium antagonists in several cerebral disorders, particularly in the syndrome of "cerebral vasospasm" after subarachnoid haemorrhage and in cerebral ischaemia associated with acute vessel occlusion where the clinical use of calcium antagonists so far has been promising.

Site of Action of Nimodipine

Calcium antagonists thus seem to have a preferential action on cerebral and coronary vessels[5]. Nimodipine shares this relatively selective effect, but it is not established whether the drug is more selective for cerebral structures than other calcium antagonists. Nevertheless, apparently based on the assumption that this is the case, nimodipine has become the most widely used drug for treatment of various disorders of the central nervous system.

In isolated feline cerebral arteries, nimodipine had a certain degree of selectivity for cerebral vessels compared to nifedipine, verapamil and diltiazem[4]. On the other hand, Brandt *et al.*[13] found the relaxant potency of nimodipine to be similar in K^+-contracted isolated human pial and mesenteric arteries. However, the high lipid solubility of nimodipine (compared with *e.g.*, nifedipine) may lead to accumulation and high concentrations of the drug in cerebral tissue. This may contribute to a preferential cerebral effect.

Beside the well documented vascular actions *in vitro* as well as *in vivo*, the beneficial effects of nimodipine in ischaemia may point to the possibility that in addition to its cerebrovascular site of action, nimodipine may also have direct neuronal effects (see below). The binding of nimodipine to receptors in rat brain was studied by Belleman *et al.*[9]. Nimodipine exhibited reversible and saturable binding to partially purified brain membranes. The nimodipine receptor was found to be highly specific and stereoselective for the dihydropyridines with calcium antagonistic action. The receptors for 1,4-dihydropyridines are also present in human brain. Peroutka and Allen[74] found that nimodipine binds to human brain membranes with a dissociation constant similar to that obtained for guinea pig ileum or rat heart muscle membranes.

Heffez and coworkers[40] investigated the nimodipine binding to gerbil brain using tritiated (H_3)-nimodipine as the radioactive ligand. Nimodipine could be detected in the brains of animals sacrificed soon after drug injection and reached a peak level within 15 minutes. Brain drug level at a given time was a linear function of the dose administered. It was concluded that in this species, effective tissue nimodipine levels may be achieved, at doses which minimize the risk of systemic hypotension.

Van den Kerkhoff and Drewes[51] estimated the transfer kinetics of nimodipine, and – in comparison – of nifedipine across the blood brain barrier (BBB) and into the cerebrospinal fluid, using the extracellular marker sucrose as a reference. The transfer into CSF was low for both drugs. However, the influx constant to the brain was much higher for nimodipine than for nifedipine. Also the regional distributions differed markedly. Nimodipine exhibited a distinctly preferential distribution to the grey matter and concentrated in different structures of the brain that are known to bear high concentrations of binding sites for dihydropyridines. The authors concluded that the high transfer of nimodipine across the BBB as well as the specific regional distribution may explain the preferential cerebral therapeutic effectiveness of the drug.

Cerebral Vasospasm

The cause of delayed cerebral vasospasm after aneurysmal subarachnoid haemorrhage (SAH) remains obscure and the significance of the various mechanisms possibly involved in this phenomenon remains controversial. A relationship between the amount and distribution of subarachnoid blood detected by computerized tomography (CT) to the later development of cerebral vasospasm in patients with ruptured saccular aneurysms was reported by Fischer and co-workers[25]. The authors found that when subarachnoid blood was not detected or was distributed diffusely on CT, severe vasospasm was almost never encountered. In the presence of subarachnoid blood clots larger than 5 × 3 mm or layers of blood 1 mm or more thick in fissures and vertical cisterns, severe spasm followed almost invariably. There was also a close correlation

between the site of the major subarachnoid blood clots and the location of severe vasospasm.

A large number of different substances occurring in posthaemorrhagic CSF from patients with ruptured aneurysms have been suggested to account for the pathogenesis of delayed cerebral hypoperfusion and the list has grown as new vasoactive agents have been discovered[101]. Among the substances incriminated in cerebral vasospasm are arenaline, NA, 5-HT, angiotensin II, prostanoids, haemoglobin, potassium and others (see[42, 47, 90, 98]). However, none has been shown to be more important than the others, and no antagonist of a single mediator candidate has been demonstrated to be therapeutically effective[101].

The experimental findings on isolated vessels that calcium antagonists, such as nimodipine, effectively prevent contractions of isolated human cerebral arteries, induced by almost all contractile agents including blood and posthaemorrhagic cerebrospinal fluid[14, 15, 22, 67, 80] made it logical to propose the use of such drugs in the prevention or treatment of cerebral vasospasm[2, 17, 99].

Further support for such use was obtained in recent *in vivo* experiments in dogs subjected to experimental SAH by a cisternal injection of autogenous blood[28]. Intrathecal administration of 4 ml of 10^{-3} M nimodipine promptly and completely reversed vasospasm in all animals studied. However, conflicting experimental results have been obtained by others[23]. Although administration of calcium antagonists may be less effective once vasospasm has developed in humans[32], there is accumulating evidence that such treatment when started early after haemorrhage may prevent or reduce the occurrence of later symptomatic vasospasm and secondary cerebral ischaemia[1, 6, 21, 50, 52, 55, 59, 68, 75, 105]. The clinical results in patients are in some contrast to some animal studies on experimental subarachnoid haemorrhage where nimodipine has failed to counteract the development of angiographically visible vasospasm. In a series of experiments on experimental subarachnoid blood deposition in monkeys, Weir and coworkers were neither able to demonstrate any beneficial clinical effect of nimodipine, nor any significant effect on angiographically visible vasospasm[24, 53, 71, 72]. The reason for this discrepancy may be explained by the fact that delayed ischaemic deterioration (DID) after SAH in humans with a ruptured intracranial aneurysm appears to be more or less an exclusively human syndrome. At least there is so far no well-controlled reproducible animal model available which mimics the human situation. In most clinical studies using intravenous nimodipine in aneurysmal SAH, the incidence of DID leading to permanent deficits or death has been decreased from an average of approximately 15–20% to less than 5%. Most of these pioneering studies have been open, non-randomized and lacking placebo controls. In spite of the very marked reduction of the incidence of DID the results of these studies have consequently been considered controversial. However, recent data from a well-designed, prospective, placebo-controlled clinical trial offer strong support for a reduction of secondary ischaemic dysfunction and also in mortality in nimodipine-treated SAH patients[105].

In addition to the actions on smooth muscle, it has been suggested that calcium-antagonists may have other effects, *e.g.* on platelet functions which may be of importance. Even if the role of platelets on the development of DID is unknown it can not be excluded that the vessel injury or tissue ischaemia may activate platelets to release potent vasoconstrictors and pro-aggregatory substances such as thromboxane and 5-HT, and this may in turn cause cerebrovascular injury. A recent study, however[103], did not support the idea that nimodipine exerts its beneficial effects in SAH patients by actions on platelet functions.

A pharmacokinetic study revealed that with the dosage used (approximately 2 mg/h), a steady state concentration of 26.6 ± 1.8 ng/ml was reached during intravenous infusion[102]. The drug was well tolerated and no serious side effects were documented.

Postischaemic Cerebral Hypoperfusion and Cerebral Ischaemia

Ischaemic brain damage is a complex phenomenon involving several pathogenetic factors[84]. Restoration of the blood supply after more than ten minutes of global ischaemia has been shown to produce functional neuronal recovery, which was again suppressed in the later phase of the postischaemic period[45]. Several observations indicate that an increased vascular resistance to flow may be a decisive factor behind reperfusion-induced tissue damage. Ames *et al.*[3] suggested localized vascular changes to be responsible for the fact that when blood flow to the brain is restored after prolonged ischaemia some areas are not reperfused and ultimately die. Ischaemic vascular damage of the endothelium may lead to loss of an endothelial derived relaxant factor (EDRF)[26,96], with subsequent increase in vascular tone. In accordance it was found that the pial artery response to *e.g.* NA is potentiated by endothelium removal[82]. Neither intravascular clotting, nor platelet aggregates

appear to cause postischaemic vascular obstruction but a contractile state of the cerebral resistance vessels may be responsible for the reperfusion damage. In cats, focal ischaemia was produced by clamping the middle cerebral artery (MCA)[15]; in initial vasodilation was observed followed by a marked, longstanding vasoconstriction. Topical application of nifedipine immediately reversed the vasoconstriction. A direct relationship between the vasoconstriction and an increased extravascular K^+-concentration was later demonstrated in the same model by Teasdale *et al.*[92]. Kazda and Mayer[49] produced seven minutes of global ischaemia in cats and found an increase of the extracellular K^+-concentration at the surface of the temporal cortex to values of more than 50 mmol/l. It was suggested that the high extracellular K^+-concentration produced sustained vascular contraction by depolarization of the vessels. The importance of increased extracellular potassium in cerebral ischaemia was emphasized by Brandt *et al.*[14] who found that the threshold for K^+-induced contraction in isolated human pial arteries was below 10 mmol/l, to be compared to the threshold in *e.g.*, mesenteric arteries, where it was 5–15 mM higher[14].

K^+ is considered to induce vascular contraction by stimulating calcium influx through potential calcium channels. Such K^+-induced vascular constriction may be one of the factors leading to an aggravated ischaemic insult. Another might be that an increased extracellular K^+ concentration enhances calcium influx into the neurones thereby aggravating the ischaemic damage. The effects of dihydropyridine drugs on the voltage sensitive influx of Ca^{++} into central nervous system neurones grown in primary culture were investigated by Thayer *et al.*[91]. They found that the depolarization induced by increasing extracellular K^+ was associated with neuronal calcium influx, which was inhibited by calcium antagonists. In a recent study by Grotta *et al.*[33] it was showed that dihydropyridines are available to binding sites and calcium channels in rat neurones.

If the assumption is true, that K^+-induced neuronal calcium influx through voltage-operated calcium channels play a role in ischaemic brain damage, then it should be expected that calcium antagonists, which effectively block these calcium channels, exert a therapeutic anti-ischaemic effect.

Global Ischaemia

In dogs subjected to 10 minutes of complete ischaemia by temporary ligation of the aorta, nimodipine improved neurological recovery and nearly doubled cerebral blood flow in the delayed postischaemic hyperperfusion period without significant effects on brain metabolism[88]. Four out of five nimodipine treated dogs were normal whereas six of seven controls were either severely damaged or dead. Pretreatment with nimodipine in cats subjected to global ischaemia did not reduce the postischaemic increase in extracellular K^+, but completely prevented the subsequent reduction of cerebral blood flow[49]. In another study nimodipine was given randomly soon after 17 minutes of complete cerebral ischaemia in monkeys[89]. After 96 hours postischaemia the neurological outcome was significantly better in the nimodipine treated animals compared to the controls. Histopathological examination yielded a significantly better mean score for the nimodipine treated group. It was concluded that nimodipine improves the neurological outcome when given after an episode of complete cerebral ischaemia. In a later study in dogs from the same laboratory, it was found that even when nimodipine treatment was delayed up to 60 minutes after reperfusion, CBF increased and the outcome was improved[60]. Another calcium channel antagonist, flunarizine, given under similar conditions did not improve neither cerebral blood flow nor neurologic outcome[70].

Vibulsresth *et al.*[100] subjected rats to a 20 minute period of ischaemia and found no beneficial effect of nimodipine when given three minutes after restoration of circulation to the brain. Conflicting results were reported by Mabe *et al.*[56]. They produced severe forebrain ischaemia in rats by four vessel occlusion with mild hypotension. After 30 minutes of ischaemia, recirculation was started by removal of the arterial clamps and by increasing blood pressure to the preischaemic level. Recovery of EEG activity following recirculation was better in the nimodipine-treated than in the control group. At two hours following recirculation, recovery of the ATP-level was also significantly better in the treated group. The promotion of functional and metabolic recovery was attributed to either improvement of post-ischaemic hypoperfusion or a direct action on metabolic processes during the reperfusion period.

Heffez and Passonneau[40] studied the effect of nimodipine on cerebral metabolism during ischaemia and post-ischaemic reflow in female mongolian gerbils. Cerebral ischaemia was induced by bilateral common carotid artery occlusion for 1, 2, or 5 minutes. After recirculation, regional levels of metabolites were measured. Pretreatment with nimodipine retarded the fall in ATP and facilitated the recovery of glucose. Regional variability was observed. The authors stated that the beneficial effects of nimodipine in experimental cerebral ischaemia, reflects also metabolic effects and should not only be attributed to effects on cerebral blood flow.

Focal Ischaemia

Gotoh *et al.*[30] administered nimodipine five minutes after MCA occlusion in the rat. The drug neither modified the pattern of cerebral blood flow distribution after occlusion, nor the extent of ischaemic brain damage as determined by histological examination. The effects of pretreatment with nimodipine were studied by Mohamed *et al.*[64] using the same autoradiography technique to registere local cerebral blood flow (lCBF) in rats subjected to MCA occlusion. Untreated control animals showed profound localized reductions in CBF 30 minutes after MCA occlusion whereas in animals pretreated with nimodipine 30 minutes before and 30 minutes after MCA occlusion the ipsilateral decrease in local CBF in cortical regions was significantly less than that in control animals. Neuropathological quantification of the ischaemic damage presented 3 hours after occlusion showed that nimodipine pretreatment reduced the volume and extent of cellular damage in the periphery but not in the core of the lesion.

In a recent study, Germano *et al.*[27], also in a rat model, described the effect of nimodipine on cerebral ischaemia in animals subjected to 1, 4, or 6 hours of MCA occlusion. The results showed that nimodipine improved neurological outcome and also decreased the infarct size when administered up to 6 hours after the ischaemic insult.

Meyer and coworkers[61] reported the effect of i.v. nimodipine on intracellular brain pH, cortical blood flow and EEG in experimental focal cerebral ischaemia in rabbits. Ischaemia was induced by MCA occlusion. In animals given nimodipine after MCA occlusion, blood flow increased and there was an associated elevation of intracellular brain pH. Cortical inspection also revealed reversal of cortical pallor and small vessel spasm following treatment with nimodipine. The authors hypothesized that nimodipine exerts its effects through reversal of postischaemic secondary vasoconstriction.

Although some experimental studies appear conflicting as to whether nimodipine is effective when given after the restoration of cerebral blood flow following an ischaemic insult, the possibility of such a beneficial effect in man is suggested by the open study by Roine *et al.*[79]. Out of 19 patients resuscitated after ventricular fibrillation and receiving intravenous nimodipine, 14 survived and 12 could be discharged home whereas in a historical control group of 19 patients, 14 died and only 5 survived and could be discharged home.

Neuronal Effects of Nimodipine

Cohen and McCarty[20] studied the nimodipine block of calcium channels in rat anterior pituitary cells. They found that dihydropyridines can block calcium channel currents in endocrine cells with very high affinity, but only during maintained depolarizations and that not all calcium channels seem to be susceptable to high affinity nimodipine block. A model was formulated and tested that predicts the amount of nimodipine binding as a function of time, voltage and drug concentration. The authors discussed the existing parallel between the block of calcium channels by nimodipine and the block of sodium channels by lidocaine and suggested that nimodipine may specifically suppress calcium entry in ischaemic tissue, with little effect on healthy, well polarized cells. During cerebral ischaemia or spreading cortical depression, extracellular K^+ can rapidly increase to 50 mM or more, thereby causing calcium entry during long-lasting depolarizations[35]. Nimodipine is likely to potently block calcium entry under such pathological conditions, whereas calcium entry under normal physiological conditions does not appear to be affected[87].

Epilepsy

A critical factor in the genesis of epileptic seizures might be neuronal calcium influx. Current investigations suggest that intrinsic neuronal bursting is dependent on calcium transport into the neurone[76, 77, 78]. It has been suggested that anticonvulsants such as phenytoin, barbiturates and benzodiazepines may act in part by preventing calcium influx at presynaptic terminals[29].

Meyer *et al.*[62] investigated the influence of nimodipine in rabbits with seizures, induced by either ischaemia, postischaemic reperfusion, pentylenetetrazol, or bicuculline. In 30 animals subjected to four hours of ischaemia, 9 of the 15 control animals had seizures in comparison with 1 of 15 nimodipine treated animals. In animals with reperfusion seizures similar results were obtained. In 10 animals in which a convulsant was applied topically to both cerebral hemispheres, unilateral intracarotid injection of nimodipine arrested seizures in that hemisphere alone, whereas the control contralateral hemisphere continued to have electrical seizure activity. Both placebo and verapamil were in-

effective. Similar effects of nimodipine on cefalozolin-induced epileptogenic foci, also in rabbits, were described by Morocutti *et al.*[66] The findings suggest that calcium influx is a common biochemical precipitant for various types of experimental seizures and that selective central nervous system calcium channel blockers may prove to be a new class of anticonvulsant agents.

Withdrawal Syndromes

Little and coworkers[54] tested the effects of four calcium channel antagonists on the ethanol withdrawal syndrome in rats. Withdrawal from chronic ethanol intake results in a syndrome of tremor and hyperexcitability which can progress to seizures and death. Nimodipine as well as nitrendipine abolished all spontaneous seizures and prevented or reduced seizures following on audiogenic stimulus and also reduced mortality. The dihydropyridines proved to be considerably more effective than diazepam in the withdrawal syndrome. These results provide evidence that changes in calcium conductance may be involved in the ethanol withdrawal syndrome and offer possibilities for the development of non-sedative therapeutic treatment of the ethanol withdrawal syndrome.

In experiments on rats, Bongianni *et al.*[12] found that nimodipine and verapamil dose-dependently reduced most of the signs of morphine abstinence. Nimodipine pretreatment markedly reduced the changes in noradrenaline content induced by the abstinence syndrome. The results suggest that calcium antagonists suppress the behavioural and neurochemical expressions of morphine abstinence by a mechanism that differs from those of opoids or α_2-adrenoceptor agonists.

Anti-nociceptive Actions

Experimental evidence from studies by Cardenas and Ross[18] and Harris *et al.*[36] of a close relationship between opiate effects and calcium transport through membranes of the central nervous system neurones and the demonstration by Chapman and Wayne[19] that morphine inhibits calciums ion influx in neuronal cells inspired Hoffmeister and Tettenborn[44] to assess the possible antinociceptive effect of nimodipine. The authors further evaluated whether and to what extent this compound interferes with the anti-nociceptive actions of an opiat-μ-receptor against (fentanyl) on a variety of pain-related behavioural reactions of the rat and the mouse. The results were correlated with general CNS effects as well as with cardiovascular effects. The authors found that the potency of nimodipine in potentiation of fentanyl anti-nociception correlated with its relative potency as calcium antagonist as measured by receptor binding studies, effects on vascular and cardiac muscle and with its neuropharmacological action (anticonvulsive effects, inhibition of balance and spontaneous motility as well as tranquilizing effects in the mouse). It was concluded that nimodipine potentiated α-receptor agonist induced anti-nociceptive effects.

References

1. Allen GS, Ahn HS, Preziosi TJ, Battye R, Boone SC, Chou SN, Kelly DL, Weir BK, Crabbe RA, Lavik PJ, Rosenbloom SB, Dorsey FC, Ingram CR, Mellits DE, Bertsch LA, Boisvert DPJ, Hundley MB, Johnson RK, Strom JoA, Transou CR (1983) Cerebral arterial spasm – a controlled trial of nimodipine in patients with subarachnoid haemorrhage. N Engl J Med 308: 619–624

2. Allen GS, Banghart SB (1979) Cerebral arterial spasm 9. *In vitro* effects of nifedipine on serotonin – phenylephrine – and potassium-induced contractions of canine basilar and femoral arteries. Neurosurgery 4: 37–42

3. Ames A, Wright RL, Kodawa M, Thurston SM, Majno G (1968) Cerebral ischaemia II. The no-reflow phenomenon. Am J Pathol 52: 437–453

4. Andersson KE, Edvinsson L, MacKenzie ET, Skärby T, Young AR (1983) Influence of extracellular calcium and calcium antagonists on contractions induced by potassium and prostaglandin F_2 in isolated cerebral and mesenteric arteries of the cat. Br J Pharmac 79: 135–140

5. Andersson KE (1986) Pharmacodynamic profiles of different calcium channel blockers. Acta Pharmacol Toxicol 58: II, 31–42

6. Auer LM (1984) Acute operation and preventive nimodipine improve outcome in patients with ruptured cerebral aneurysms. Neurosurgery 15: 57–66

7. Auer LM, Mokry M (1985) Effect of nimodipine and its solvent on superficial cerebral vessels. J Cereb Blood Flow Metab 5: 473–474

8. Auer LM, Mokry M (1986) Effect of topical nimodipine versus its ethanol-containing vehicle on cat pial arteries. Stroke 17: 225–228

9. Belleman P, Schade A, Towart R (1983) Dihydropyridine receptor in rat brain labelled with ^{3}H-nimodipine. Proc Natl Acad Sci USA 80: 2356–2360

10. Bevan JA (1982) Selective action of diltiazem on cerebral vascular smooth muscle in the rabbit: antagonism of extrinsic but not intrinsic maintained tone. Am J Cardiol 49: 519–524

11. Bolton TB (1979) Mechanisms of action of transmitters and other substances on smooth muscle. Pharmacol Rev 59: 67–72

12. Bongianni F, Carla V, Moroni F, Pellegrini-Giampietro DE (1986) Calcium channel inhibitors suppress the morphine-withdrawal syndrome in rats. Br J Pharmac 88: 561–567

13. Brandt L, Andersson KE, Edvinsson L, Ljunggren B (1981) Effects of extracellular calcium and calcium antagonists on the contractile responses of isolated human pial and mesenteric arteries. J Cereb Blood Flow Metab 1: 339–347

14. Brandt L (1981) Aspects on cerebral vasospasm. A clinical and experimental study. Doctoral Thesis, University of Lund, Sweden

15. Brandt L, Ljunggren B, Andersson KE, MacKenzie ET, Tamura A, Teasdale G (1982) Effects on feline cortical pial microvasculature of topical application of a calcium antagonist (nifedipine) under normal conditions and in focal ischaemia. J Cereb Blood Flow Metab 3: 44–50

16. Brandt L, Ljunggren B, Säveland H (1986) Prophylaxe ischämischer neurologischer Defizite nach Subarachnoidalblutung. Kombinierte Behandlung aus Frühoperationen des Aneurysmas und Nimodipin. In: Kazner E (ed) Krankenhausarzt. Braun Verlag Medizinische Bücher, GmbH, Karlsruhe, pp 59–66

17. Brandt L, Andersson KE, Bengtsson B, Edvinsson L, Ljunggren B, MacKenzie ET (1979) Effects of nifedipine on pial arteriolar calibre: an *in vivo* study. Surg Neurol 12: 349–352

18. Cardenas HL, Ross DH (1975) Morphine induced calcium depletion in discrete regions of rat brain. J Neurochem 24: 487–493

19. Chapman DB, Way EL (1980) Metal ion interaction with opiates. Ann Rev Pharmacol Toxicol 20: 553

20. Cohen CJ, McCarthy RT (1987) Nimodipine block of calcium channels in rat anterior pituitary cells. J Physiol 387: 195–225

21. Disney L, Weir B (in press) Nimodipine treatment in poor grade patients. Results of a multicenter, double-blind, placebo controlled study. In: Proceedings of cerebral vasospasm 1987 – a research conference. Williams & Wilkins, Baltimore, USA

22. Edvinsson L, Brandt L, Andersson KE, Bengtsson B (1979) Effect of a calcium antagonist on experimental constriction of human brain vessels: possible efficacy in cerebrovascular spasm. Surg Neurol 11: 327–330

23. Espinosa F, Weir B, Overton T, Castor W, Grace M, Boisvert D (1984) A randomized placebo-controlled double-blind trial for nimodipine after SAH in monkeys. Part 1: Clinical and radiological findings. J Neurosurg 60: 1167–1175

24. Espinosa F, Weir B, Shnitka T, Overton T, Boisvert D (1984) A randomized placebo-controlled double-blind trial for nimodipine after SAH in monkeys. Part 2: Pathological findings. J Neurosurg 60: 1176–1185

25. Fisher CM, Kistler JP, Davis JM (1980) Relation of cerebral vasospasm to subarachnoid haemorrhage visualized by computerized tomographic scanning. Neurosurgery 6: 1–9

26. Furchgott RF (1984) The role of endothelium in the responses of vascular smooth muscle to drugs. Ann Rev Pharmacol Toxicol 24: 175–197

27. Germano IM, Bartkowski HM, Cassel ME, Pitts LH (1987) The therapeutic value of nimodipine in experimental focal cerebral ischaemia. Neurological outcome and histopathological findings. J Neurosurg 67: 81–87

28. Gioia AE, White RP, Bakiitian B, Robertson JT (1985) Evaluation of the efficacy of intrathecal nimodipine in canine models of chronic cerebral vasospasm. J Neurosurg 62: 721–728

29. Glaser GH, Penry JK, Woodbury DM (1980) Antiepileptic drugs: mechanisms of action. Advances in neurology, vol 27. Raven Press, New York

30. Gotoh O, Mohamed OA, McCulloch J, Graham DI, Harper AM, Teasdale GM (1986) Nimodipine and the haemodynamic and histopathological consequences of middle cerebral artery occlusion in the rat. J Cereb Blood Flow Metab 6: 321–331

31. Greenberg DA (1987) Calcium channels and calcium channel antagonists. Ann Neurol 21: 317–330

32. Grotenhuis JA, Bettag W, Othmar Fiebach BJ, Dabir K (1984) Intracarotid slow bolus injection of nimodipine during angiography for treatment of cerebral vasospasm after SAH. A preliminary report. J Neurosurg 61: 231–240

33. Grotta JC, Creed Pettigrew L, Lockwood AH, Reich C (1987) Brain extraction of a calcium channel blocker. Ann Neurol 21: 171–175

34. Hagiwara S, Byerly L (1981) Calcium channel. Ann Rev Neurosci 4: 69–125

35. Hansen AJ (1985) Effect of anoxia on ion distribution in the brain. Physiol Rev 65: 101–148

36. Harris RA, Yamamoto H, Loh HH, Way EL (1977) Discrete changes in brain calcium with morphine analgesia, tolerance, dependence, and abstinence. Life Sci 20: 501–506

37. Havanka-Kanniainen H, Hokkanen E, Myllylä VV (1987) Efficacy of nimodipine in comparison with pizotifen in the prophylaxis of migraine. Cephalalgia 7: 7–13

38. Haws CW, Gourley JK, Heistad DD (1983) Effects of nimodipine on cerebral blood flow. J Pharmacol Exp Ter 225: 1, 24–28

39. Hayashi S, Toda N (1977) Inhibition by Ca^{2+}, verapamil, and papaverine of Ca^{2+}-induced contractions in isolated cerebral and peripheral arteries of the dog. Br J Pharmacol 60: 35–43

40. Heffez Ds, Passonneau JV (1985) Effect of nimodipine on cerebral metabolism during ischaemia and recirculation in the mongolian gerbil. J Cereb Blood Flow Metab 5: 523–528

41. Heffez DS, Nowak Jr TS, Passonneau JV (1985) Nimodipine levels in gerbil brain following parenteral drug administration. J Neurosurg 63: 589–592

42. Heros RC, Zervas NT, Varsos V (1983) Cerebral vasospasm after subarachnoid haemorrhage: an update. Ann Neurol 14: 599–608

43. Hof RP (1983) Calcium antagonists and the peripheral circulation: differences and similarities between PY 108-068, nicardipine, verapamil and diltiazem. Br J Pharmacol 78: 375–394

44. Hoffmeister F, Tettenborn D (1986) Calcium agonists and antagonists of the dihydropyridine type: Antinociceptive effects, interference with opiate-u-receptor agonists and neuropharmacological actions in rodents. Psychopharmacology 90: 299–307

45. Hossman KA, Sato K (1970) Recovery of neuronal function after prolonged cerebral ischaemia. Science 168: 375–376

46. Högestätt ED, Andersson KE, Edvinsson L (1982) Effects of nifedipine on potassium-induced contraction and noradrenaline release in cerebral and extracranial arteries from rabbit. Acta Physiol Scand 114: 283–296

47. Kassell NF, Sasaki T, Colohan ART, Nazar G (1985) Cerebral vasospasm following aneurysmal subarachnoid haemorrhage. Stroke 16: 562–572

48. Kazda S, Garthoff B, Krause HP, Schlossman K (1982) Cerebrovascular effects of the calcium antagonistic dihydropyridine derivative nimodipine in animal experiments. Arzneimittelforschung (Drug Res) 32: 331–338

49. Kazda S, Mayer D (1985) Postischaemic impaired reperfusion and tissue damage: consequences of a calcium dependent vasospasm. In: Goodfraind T, Vanhoutte PM, Govoni S, Paoletti R (eds) Calcium entry blockers and tissue protection. Raven Press, New York, pp 129–149

50. Kazner E, Sprung CH, Adelt D *et al* (1985) Clinical experience with nimodipine in the prophylaxis of neurological deficits after subarachnoid haemorrhage. Neurochirurgia 28: 110–113

51. van den Kerkhoff W, Drewes LR (1985) Transfer of the Ca-antagonists nifedipine and nimodipine across the blood-brain barrier and their regional distribution *in vivo*. J Cereb Blood Flow Metab 5: 1, 459–460

52. Koos WT, Perneczky A, Auer LM *et al* (1985) Nimodipine treatment of ischaemic neurologic deficits due to cerebral vasospasm after subarachnoid haemorrhage. Neurochirurgia 28: 114–117

53. Krueger C, Weir B, Nosko M, Cook D, Norris S (1985) Nimodipine and chronic vasospasm in monkeys: Part 2. Pharmacological studies of vessels in spasm. Neurosurgery 16: 2, 137–140

54. Little HJ, Dolin SJ, Halsey MJ (1986) Calcium channel antagonists decrease the ethanol withdrawal syndrome. Life Sciences 39: 22, 2059–2065

55. Ljunggren B, Brandt L, Säveland H, Nilsson PE, Cronquist S, Andersson KE, Vinge E (1984) Outcome in 60 consecutive patients treated with early aneurysm operation and intravenous nimodipine. J Neurosurg 61: 864–873

56. Mabe H, Nagai H, Takagi T, Umemura S, Ohno M (1986) Effect of nimodipine on cerebral functional and metabolic recovery following ischaemia in the rat brain. Stroke 17: 3, 501–505

57. McCalden TA, Bevan JA (1981) Sources of activator calcium in rabbit basilar artery. Am J Physiol 241: H 129–H 133

58. McCalden TA, Nath RG, Thiele K (1984) The effects of a calcium antagonist (nimodipine) on basal cerebral blood flow and reactivity to various agonists. Stroke 15: 527–530

59. Mee EW, Dorrance DE, Low D, Neil-Dwyer G (1986) Cerebral blood flow and neurological outcome: a controlled study of nimodipine in patients with subarachnoid haemorrhage. J Neurol Neurosurg Psychiatry 49: 469

60. Milde LN, Milde JH, Michenfelder JD (1986) Delayed treatment with nimodipine improves cerebral blood flow after complete cerebral ischaemia in the dog. J Cereb Blood Flow Metab 6: 332–337

61. Meyer FB, Anderson RE, Yaksh TL, Sundt TM (1986) Effect of nimodipine on intracellular brain pH, cortical blood flow, and EEG in experimental focal cerebral ischaemia. J Neurosurg 64: 617–626

62. Meyer FB, Anderson RE, Sundt TM, Sharbrough FW (1986) Selective central nervous system calcium channel blockers – a new class of anticonvulsant agents. Mayo Clin Proc 61: 239–247

63. Middlemiss DN, Spedding M (1985) A functional correlate for the dihydropyridine binding site in rat brain. Nature 314: 94–96

64. Mohamed AA, Gotoh O, Graham DI, Osborne KA, McCulloch J, Mendelow AD, Teasdale GM, Harper AM (1985) Effect of pretreatment with the calcium antagonist nimodipine on local cerebral blood flow and histopathology after middle cerebral artery occlusion. Ann Neurol 18: 705–711

65. Mohamed AA, Mendelow AD, Teasdale GM, Harper AM, McCulloch J (1985) Effect of the calcium antagonist nimodipine on local cerebral blood flow and metabolic coupling. J Cereb Blood Flow Metab 5: 26–33

66. Morocutti C, Pierelli F, Sanarelli L, Stefano E, Peppe A, Mattioli GL (1986) Antiepileptic effects of a calcium antagonist (nimodipine) on cefazolin-induced epileptogenic foci in rabbits. Epilepsia 27: 498–503

67. Müller-Schweinitzer E, Neumann P (1983) *In vitro* effects of calcium antagonists PN 200-110, nifedipine, and nimodipine on human and canine cerebral arteries. J Cereb Blood Flow Metab 3: 354–361

68. Neil-Dwyer GA (1985) A controlled study of nimodipine in subarachnoid haemorrhage patients. In: Proceedings of 13th World Congress of Neurology, Hamburg, Sept 1–6

69. Newberg Milde L, Milde JH, Michenfelder JD (1986) Delayed treatment with nimodipine improves cerebral blood flow after complete cerebral ischaemia in the dog. J Cereb Blood Flow Metab 6: 332–337

70. Newberg LA, Steen PA, Milde JH, Michenfelder JD (1984) Failure of flunarizine to improve cerebral blood flow of neurologic recovery in a canine model of complete cerebral ischaemia. Stroke 15: 4, 666–671

71. Nosko M, Weir B, Krueger C, Cook D, Norris S, Overton T, Boisvert D (1985) Nimodipine and chronic vasospasm in monkeys: Part 1. Clinical and radiological findings. Neurosurgery 16: 2, 129–136

72. Nosko M, Krueger A, Weir BKA, Cook DA (1986) Effects of nimodipine on *in vitro* contractility of cerebral arteries of dog, monkey and man. J Neurosurg 65: 376–381

73. Nowycky MC, Fox AP, Tsien RW (1985) Three types of neuronal calcium channel with different calcium agonist sensitivity. Nature 316: 440–443

74. Peroutka SJ, Allen GS (1983) Calcium channel antagonist binding sites labelled by ^{3}H-nimodipine in human brain. J Neurosurg 59: 933–937

75. Philippon J, Grob R, Dagreou F, Guggiari M, Rivierez M, Viars P (1986) Prevention of vasospasm in subarachnoid haemorrhage. A controlled study with nimodipine. Acta Neurochir (Wien) 82: 110–114

76. Prince DA, Connors BW (1984) Mechanisms of epileptogenesis in cortical structures. Ann Neurol 16: 559–564

77. Prince DA (1985) Physiological mechanisms of focal epileptogenesis. Epilepsia 26: S 13–S 14

78. Pumain R, Kurcewicz I, Louvel J (1983) Fast extracellular calcium transients: involvement in epileptic processes. Science 222: 177–179

79. Roine RO, Kaste M, Kinnunen A, Nikki P (1987) Safety and efficacy of nimodipine in resuscitation of patients outside hospital. Br Med J 294: 20

80. Salaicies M, Maria J, Rico ML, Gonzalez C (1983) Effects of verapamil and manganese on the vasoconstrictor responses to noradrenaline, serotonin and potassium in human and goat cerebral arteries. Biochem Pharmacol 32: 2711–2714

81. Sasaki T, Kassell NF, Zuccarello M (1986) Dependence of cerebral arterial contractions on intracellularly stored Ca^{++}. Stroke 17: 95–97

82. Sercombe R, Verrechia C, Oudort N, Dimitriadon V, Seylaz J (1985) Pial artery responses to norepinephrine potentiated by endothelium removal. J Cereb Blood Flow Metab 5: 312–317

83. Shapiro HM (1985) Post-cardiac arrest therapy: calcium entry blockade and brain resuscitation. Anaesthesiology 62: 384–387

84. Siesjö BK (1984) Cerebral circulation and metabolism. J Neurosurg 60: 883–908

85. Scriabine A, Battye R, Hoffmeister F, Kazda S, Towart R, Garthoff B, Schlüter G, Rämsch KD, Scherling D (1985) Nimodipine. In: Scriabine A (ed) New drugs annual: cardiovascular drugs, vol 3. Raven Press, New York, pp 197–218

86. Skärby T, Högestätt ED, Andersson KE (1984) Influence of extracellular calcium and nifedipine on 1- and 2-adrenoceptor mediated contractile responses in isolated rat and cat cerebral and mesenteric arteries. Acta Physiol Scand 123: 445–456

87. Spedding M, Middlemiss DN (1985) Central effects of Ca^{2+} antagonists. TIPS, 309–310

88. Steen PA, Newberg LA, Milde JH, Michenfelder JD (1983) Nimodipine improves cerebral blood flow and neurologic recovery after complete cerebral ischaemia in the dog. J Cereb Blood Flow Metab 3: 38–43

89. Steen PA, Gisvold SE, Milde JH, Newberg LA, Scheithauer BW, Lanier WL, Michenfelder JD (1985) Nimodipine improves outcome when given after complete cerebral ischaemia in primates. Anesthesiology 62: 406–414

90. Sundt TM Jr, Szurszewski J, Sharbrough FW (1977) Physiological considerations important for the management of vasospasm. Surg Neurol 7: 259–267

91. Thayer SA, Murphy SN, Miller RJ (1986) Widespread distribution of dihydropyridine-sensitive calcium channels in the central nervous system. Mol Pharmacol 30: 505–509

92. Teasdale G, Legrain Y, MacKenzie E, Graham DF (1983) Potassium release and vascular events in focal cerebral ischaemia. J Cereb Blood Flow Metab 3: 1, S 395–S 396

93. Towart R (1981) The selective inhibition of serotonin-induced contractions of rabbit cerebral vascular smooth muscle by calcium antagonistic dihydropyridines. An investigation of the mechanism of action of nimodipine. Circ Res 48: 650–657

94. Towart R, Wehinger E, Meyer H, Kazda S (1982) The effects of nimodipine, its optical isomers and metabolites on isolated vascular smooth muscle. Arzneimittelforschung (Drug Res) 32: 338–346

95. Uski T, Andersson KE (1984) Effects of prostanoids on isolated feline cerebral arteries II. Roles of extra- and intracellular calcium for the prostaglandin F_2-induced contraction. Acta Physiol Scand 120: 197–205

96. Vanhoutte PM (1986) Could the absence or malfunction of vascular endothelium precipitate the occurrence of vasospasm? J Mol Cell Cardiol 18: 679–689

97. Vanhoutte PM (1987) The expert committee of the world health organization on classification of calcium antagonists: the viewpoint of the raporteur. Am J Cardiol 59: 3 A–8 A

98. White RP (1979) Multiplex origins of cerebral vasospasm. In: Price TR, Nelson E (eds) Cerebrovascular diseases. Raven Press, New York, pp 573–581

99. White RP, Cunningham MP, Robertson JT (1982) Effect of the calcium antagonist nimodipine on contractile responses of isolated canine basilar arteries induced by serotonin, prostaglandin F_2, thrombin, and whole blood. Neurosurgery 10: 3, 344–348

100. Vibulsresth S, Dietrich WD, Busto R, Ginsberg MD (1987) Failure of nimodipine to prevent ischaemic neuronal damage in rats. Stroke 18: 210–216

101. Wilkins RH (1986) Attempts at prevention or treatment of intracranial arterial spasm: an update. Neurosurgery 18: 808–825

102. Vinge E, Andersson KE, Brandt L, Ljunggren B, Nilsson LG, Rosendal-Helgesen S (1986) Pharmacokinetics of nimodipine in patients with aneurysmal subarachnoid haemorrhage. Eur J Pharmacol 30: 421–425

103. Vinge E, Brandt L, Ljunggren B, Andersson KE (1987) Thromboxane B 2 levels in serum during continuous administration of nimodipine to patients with aneurysmal SAH. Stroke, submitted

104. Yamamoto M, Ohta T, Toda N (1983) Mechanisms of relaxant action of nicardipine, a new Ca^{++}-antagonist, on isolated dog cerebral and mesenteric arteries. Stroke 14: 270–275

105. Öman J (1987) Prevention of ischaemic dysfunction in patients with aneurysmal subarachnoid haemorrhage. Presented at 2nd World Congress of Neurosciences; Session 132: Mechanism and prevention of cerebral vasospasm. Budapest, Hungary

Correspondence: L. Brandt, M.D., Department of Neurosurgery, University Hospital, S-221 85 Lund, Sweden.

Acta Neurochirurgica, Suppl. 45, 21–28 (1988)

Haemodynamic Effectiveness of Nimodipine on Spastic Brain Vessels After Subarachnoid Haemorrhage Evaluated by the Transcranial Doppler Method

A Review of Clinical Studies

A. Harders and **J. Gilsbach**

Neurochirurgische Universitätsklinik, Freiburg i. Br., Federal Republic of Germany

Summary

The authors review the literature reports and own results of a double-blind study of the effectiveness of nimodipine on prevention or treatment of spasm of cerebral arteries following subarachnoid haemorrhage (SAH). Spasm has been evaluated using the transcranial Doppler method (TCD). The patients were divided into two groups which received 2 resp. 3 mg/h nimodipine. The clinical outcome and also the incidence of spasm of both of the groups were not different, but spasm was less severe in the 3 mg/h group.

Keywords: Cerebral vasospasm; subarachnoid haemorrhage; SAH; transcranial Doppler method; TCD; nimodipine.

Introduction

The calcium channel blocker nimodipine was developed as a vasodilator for cerebral arteries. Until now, cerebral vascular spasm following subarachnoid haemorrhage (SAH) has been the major indication for nimodipine. The effect has been confirmed in animals studies[5, 13, 23, 29, 30, 31], during topical application of the drug on cerebral arteries[6, 7] and in clinical studies on patients who had suffered a subarachnoid haemorrhage[8, 9, 14–18, 27, 28, 33, 34].

Since 1982 it is possible to measure noninvasively the velocities in the large basal arteries of the brain by the transcranial Doppler (TCD) method[1, 2, 3, 19, 36]. Alterations in the diameter of the brain arteries induced by nimodipine result in blood flow velocity changes. These velocity changes can be measured by means of transcranial Doppler sonography.

Studies of TCD measurements in patients after suffering SAH who underwent surgery for cerebral aneurysm and were treated prophylactically with nimodipine are scarce.

Our own results from a dose-control double-blind study are presented. A control group of patients without nimodipine treatment was not used because clinical studies have already demonstrated the effectiveness of the drug in reducing the incidence of delayed ischaemic deficits.

Literature Review

Seiler *et al.*[37] published a prospective transcranial Doppler study in 1987. Thirty-three patients received no nimodipine, as opposed to 37 patients who were treated intravenously with 2 mg/h nimodipine. Transcranial Doppler studies were performed daily in both MCA's. Nimodipine administration was initiated within 24 hours after the SAH in most of the patients and continued for 7–14 days postoperatively. The intravenous dose was then changed to 60 mg/4 h of oral nimodipine for one week. Seiler classified the patients into low and high risk patients according to the amount of CT visualized cisternal blood. Twenty-seven patients were operated on early – within 3 days after the SAH – and 43 of the patients were operated on late. The time course of the flow velocity changes was the same in all the patients who had received nimodipine, however, statistically the velocities increased more in the non-treated patients. Low and high risk subgroups showed the same difference in nimodipine efficacy. The greatest haemodynamic effect in favour of nimodipine could be demonstrated in the high risk patients undergoing late surgery. In the low risk patients vasospasm played a minor role, however, the timing and quality of surgery were major factors for the clinical outcome. The low risk patients treated with nimodipine showed no difference in the blood flow velocities depending on whether they were operated on early or late. In the high risk group the nimodipine patients operated on also showed a significantly higher increase in blood flow velocities in comparison with those who were operated on late, but nimodipine significantly reduced the incidence of delayed ischaemic neurological deficits (DID) in this group of patients and improved their final outcome. The total incidence of symptomatic vasospasm was 30%. There was no difference in the low risk patient group, whereas in the high risk patients the incidence of (DID) was significantly reduced in the nimodipine group (p = 0.05).

Pasqualin et al.[32] in 1987 described 81 patients who had been treated with prophylactic intravenous nimodipine at a dosage of 2 mg/h. The nimodipine prophylaxis was started within 5 days post SAH and the mode of application was changed to oral administration (360 mg/d) after the second week following the haemorrhage. Fifty-one percent of the patients were in Hunt & Hess grades I–II, 34% in grade III, and 15% in grade IV on admission. Fifty-three patients were operated on within 3 days after the haemorrhage, 2 patients were operated on "subacutely", and 26 of the patients were treated later than 2 weeks after the SAH. At the onset of ischaemic deficits all patients were treated with hypervolaemia and some patients were also treated with induced hypertension. Half of the 81 patients underwent daily transcranial Doppler examination and the values were compared with 28 patients who had not been treated with nimodipine. In the 40 patients who received nimodipine and who underwent TCD examinations the incidence of pathological frequencies in the circle of Willis compared with the control group of 28 untreated cases did not seem to be significantly different. However, patients with the same amount of cisternal blood deposition (consistent/thick) and treated with nimodipine had a higher incidence of Doppler frequencies between 2 and 4 kHz and a lower incidence of values over 4 kHz (36% versus 44%) as compared with the control group (1 kHz = approx. 39 cm/sec). The incidence of ischaemic deterioration was slightly lower in patients treated with nimodipine (38% versus 47%). Nimodipine did not seem to decrease the incidence of pathological blood flow velocities in the circle of Willis. However, Pasqualin states that very high velocities (over 160 cm/sec) were observed less frequently during nimodipine administration. This is consistent with Seiler's findings[37] and those of our group which indicate that the velocity ranges during the time course of vasospasm are lowered.

Nimodipine Dose-Control Double-Blind Study (Freiburg)

Patients and Method

Forty-five patients who underwent early aneurysm surgery within 72 hours after the last haemorrhage received 2 or 3 mg/h nimodipine in a dose-control double-blind study. Transcranial Doppler measurements were performed at least every 3 days in the different segment of the circle of Willis (A 1, P 1, MCA, ICA and the syphon). To ensure that equal numbers of patients received the higher and the lower dose, the patients were classified according to the grades of Hunt & Hess[24], patients in grade I–III (good condition patients) and in patients with grade IV–V (bad condition patients). After the study had been completed, there was nearly the same number of patients in both groups, but more women than men. In the 3 mg/h group there were more patients over 60 years old than in the 2 mg/h group (Table 1). The correlation of the preoperative clinical status according to Hunt & Hess and the CT grading according to Fisher[11, 12] in the different treatment groups is shown in Tables 2 and 3. In the 3 mg group there were more patients with a severe subarachnoid haemorrhage (grade III, n = 16) than in the 2 mg group (grade III, n = 12).

Table 1. *Characteristics of the Patients*

Nimodipine dosage	N	M/F	Age (years)		
			≤ 40	41–59	≥ 60
2 mg/h	22	8/14	8	11	3
3 mg/h	23	9/14	5	9	9

Table 2. *Correlation of the Preoperative Clinical Status According to Hunt and Hess with the Source of SAH According to Fisher in Those Patients Who Received 2 mg/h Nimodipine*

2 mg/h Nimodipine (n = 22)

CT/H & H	I	II	III	IV
I	2	0	0	0
II	0	4	3	1
III	0	0	4	8

Table 3. *Correlation of the Preoperative Clinical Status According to Hunt and Hess with the Source of SAH According to Fisher in Those Patients Who Received 3 mg/h Nimodipine*

3 mg/h Nimodipine (n = 23)

CT/H & H	I	II	III	IV	V
I	1	1	0	0	0
II	0	2	3	0	0
III	0	2	7	6	1

We tried to attain normotension in our patients and to avoid hypotension to be able to achieve a sufficient brain perfusion pressure. Based on the experience in 55 patients previously treated[21], we performed a mild induced hypertension therapy with dobutamin, effortil, regulton or plasma-expander in 47% of those patients who showed a flow velocity increase to more than 120 cm/sec in the TCD measurements performed in the first week after SAH. If the Doppler measurements showed very high velocities over 160 cm/sec the patient was instructed to maintain bed rest for several days to prevent orthostasis.

Results

Using the highest measured velocity in one artery of the circle of Willis, there was a velocity increase between day 3 and 9, then a maximum plateau phase till day 21, followed by the velocities slowly returning to normal. The patients in the 3 mg group showed a less severe compensatory blood flow velocity increase. The difference peak velocities calculated by the mean velocities between the 2 groups was about 20 cm/sec (Fig. 1).

Taking only the velocities in the MCA into account, the time course was the same, the compensatory blood flow velocity increase in the 3 mg group was less than in the 2 mg group (Fig. 2).

Because the severity of vasospasm is closely related to the severity of the subarachnoid haemorrhage[12, 21, 22, 36], we compared the time course of velocity changes in patients who suffered a severe subarachnoid haemorrhage (CT = III). In both groups there was a similar

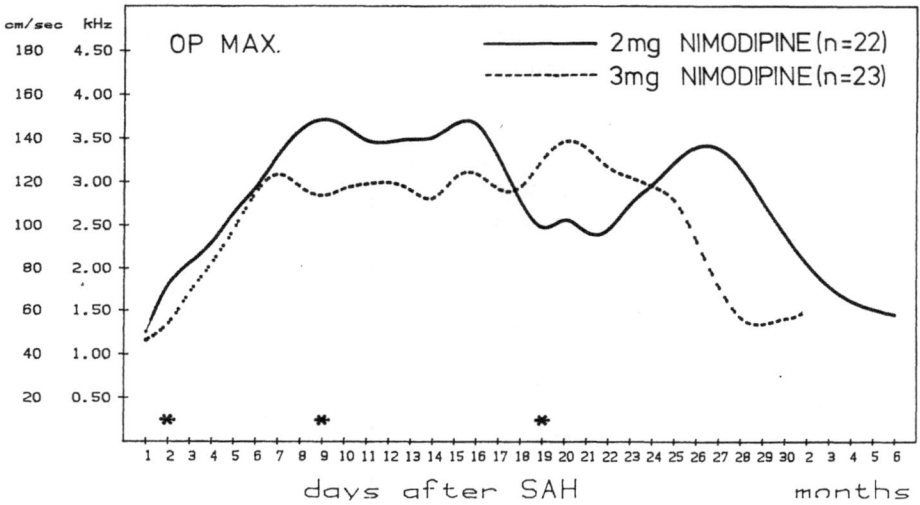

Fig. 1. Time course of the day mean velocities using the highest velocity in one artery in the two different nimodipine dosage groups of patients

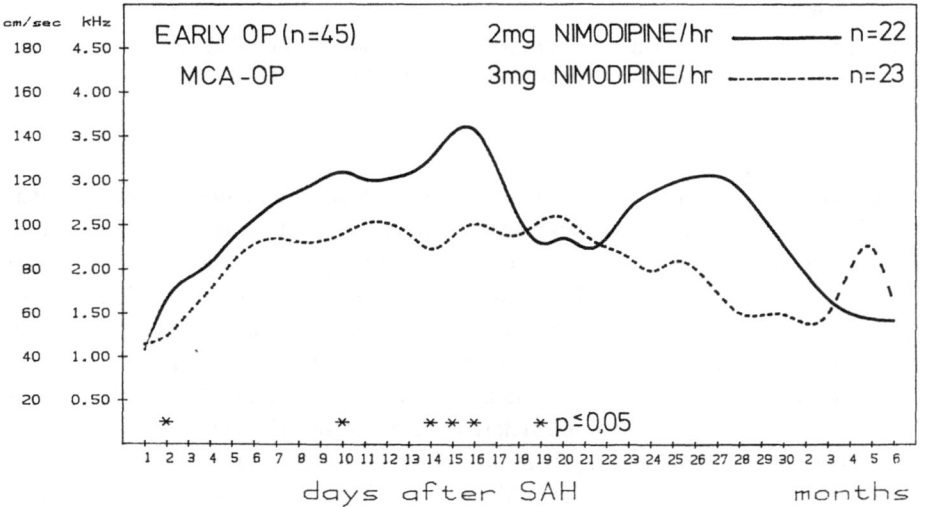

Fig. 2. Time course of day mean velocities in the MCA on the side of the operative approach in the two different nimodipine patients groups

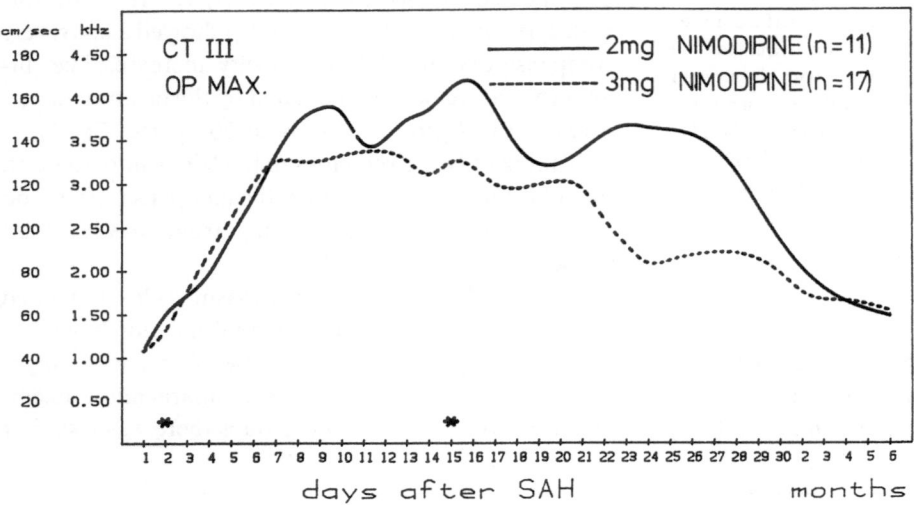

Fig. 3. Time course of day mean velocity changes of those patients who had a severe subarachnoid haemorrhage (CT = 3) in the two different nimodipine groups. The maximum velocity in one artery of the side of the operative approach (op max) was taken into account

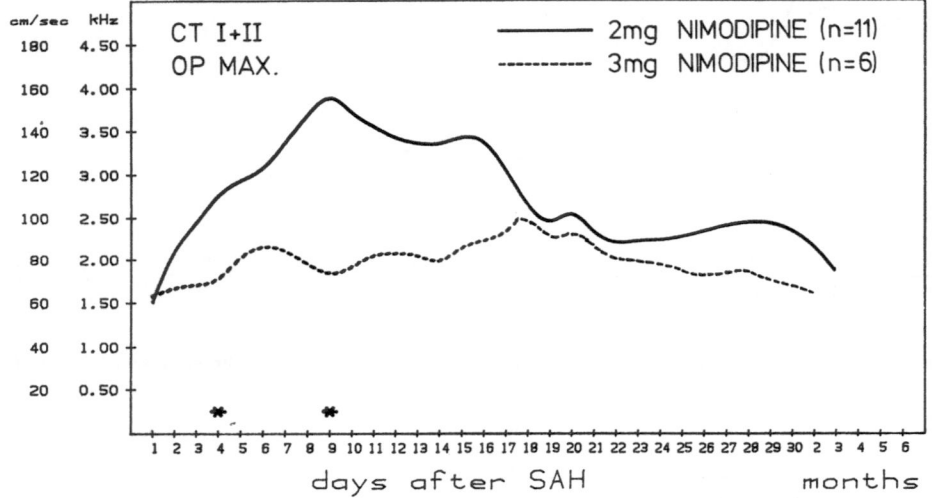

Fig. 4. Time course of day mean velocity changes in patients who had a mild or medium subarachnoid haemorrhage (CT = I or II). The maximum velocities on each day on the side of the operative approach were taken into account in the two different nimodipine groups

velocity increase in the first 6 days. However, the 3 mg group had already reached its maximum in this time while the 2 mg group showed further velocity increase (Fig. 3).

In contrast, the velocity increase in patients who suffered a mild or moderate (CT I or II) subarachnoid haemorrhage differed considerably between the two groups, as is shown in Fig. 4.

A comparison of the frequency ranges of the two groups as shown in Fig. 5 reveals that the same number of patients had no pathological frequency changes

($\leqslant 1.9$ kHz), and even the incidence of severe vasospasm was the same ($\leqslant 5.9$ kHz and $\leqslant 6.9$ kHz). However, the incidence of frequency ranges between 2 and 4.9 kHz was lower in the 3 mg group. This indicates that the incidence of vasospasm is the same in both groups, but that the severity is different.

One of the 45 patients developed a transient neurological deficit due to vasospasm. This patient received 2 mg/h nimodipine for 14 days. The clinical outcome of the patients established by the Glasgow Outcome Scale showed no difference between the two groups[16].

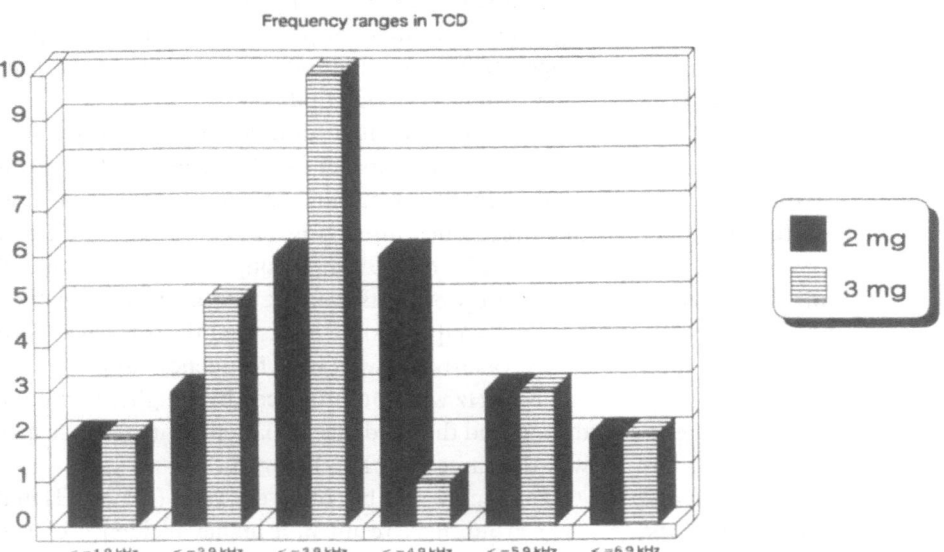

Fig. 5. The incidence of frequency ranges in the two different nimodipine groups. Up to 2 kHz it corresponds to velocities in the normal range while above 3 kHz it is called "critical" vasospasm

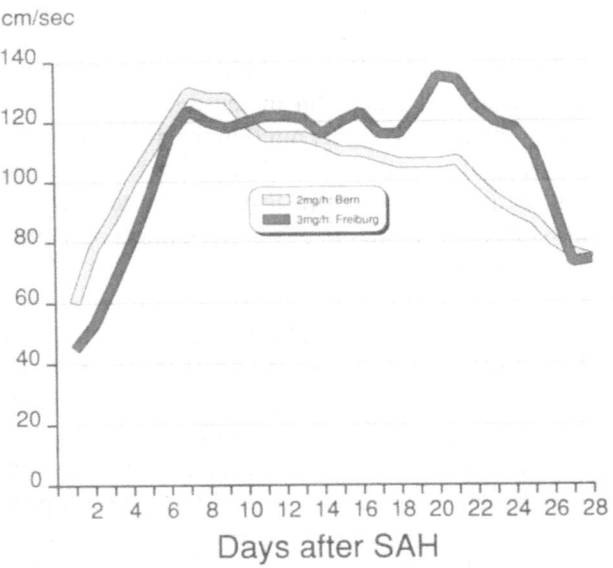

Fig. 6. Time course of the day mean velocity changes in 37 patients treated with nimodipine at a dosage of 2 mg/h (n = 37, Bern) compared with patients who had been treated with 3 mg/h nimodipine (n = 23, Freiburg)

Discussion

In the last 5 years, the diagnosis of vasospasm has become more and more reliable as a result of transcranial Doppler sonography. Due to the collateral pathways, especially in the anterior part of the circle of Willis such as the anterior communicating artery the compensatory blood flow velocity increase in the A 1 segment is less than in the middle cerebral artery or the distal part of the internal carotid artery. It could be shown that there is a statistical significant inverse relationship between the blood flow velocity and the diameter of the arteries in the MCA and ICA[2, 19, 21, 26]. When there is a velocity increase in TCD to more than 120 cm/sec, vasospasm is always demonstrated on angiography[2, 21]. Not only are angiographic studies invasive and can only be performed at least in one patient twice or three times after SAH, they mainly prove the existence of vasospasm[10, 25, 35, 39]. With the TCD method daily measurements help to evaluate the individual time course of vasospasm. The natural time course, which has been described by Seiler and Harders[19, 36], is similar in patients who underwent aneurysm surgery within 72 hours after the SAH. We know that within the first 3 days after the SAH there is no sign of vasospasm on angiography or in TCD[21]. Then a more or less rapid increase of velocities begins up to day 10–11, after which there is a maximum period which can be measured up to the end of the third week. Then the velocities slowly return to normal within the next 2 to 4 weeks. A comparison of the groups treated and not treated with nimodipine[32, 37], as well as of the groups receiving 2 different dosages of the drug described in this publication was made in order to assess the influence of the calcium channel blocker on the vessel diameter which causes velocity changes.

A very important factor is the dependency of the severity of the increase in compensatory blood flow velocity on age[22], the source of SAH[12, 21, 36], the blood pressure (hypertensive patients, hypertension therapy) — blood viscosity, haemodilution treatment, and the timing of surgery. For an evaluation of the effect of nimodipine the above parameters have to be kept as constant as possible which in clinical practice is very difficult.

In the study by Seiler *et al.*[37] patients were divided into low and high risk groups depending on the severity of the SAH. Most of the patients operated on within 3 days were in the low risk group and most of them received nimodipine (15:4). The high risk patients operated on early all received nimodipine (n = 8). The mean age of the patients was the same in both the treated and non-treated groups of high risk patients, whereas in the low risk group the mean age of the control group of non-treated subjects was lower than that of the treated patients. The velocities were high and it must be pointed out that younger patients develop more severe vasospasm than older persons, resulting in a not significant difference between the 2 curves. In the unselected patients, in the low risk group and in the high risk group, the difference in the blood flow velocities established by TCD were always statistically significantly different and in favour of the nimodipine treated patients. A comparison of time courses with those of our patients (Fig. 6) reveals an astonishing correlation up to day 14 after SAH, even though all of our patients had been operated on early. Our 3 mg/h patients show higher velocities from day 10 on than Seiler's 2 mg/h patients. That means that the initial time course of vasospasm is typically constant. However, the severity of vasospasm is influenced by the timing of the operation, the age of the patient, the severity of the SAH, and the nimodipine dosage.

Pasqualin[32] reports that about two-thirds of his patients underwent early operation, some were treated with hypervolaemia and some with induced hypertension therapy. He found lower frequency ranges under nimodipine treatment. However, the timing of surgery is not mentioned for his control group of 28 patients without nimodipine. Pasqualin concludes that prophylactic nimodipine did not seem to prevent vessel narrowing and ischaemic deterioration. He states that in contrast to Seiler and Harders, he did not find a decrease in the incidence of pathological frequencies. However, in the papers mentioned above, similar results are reported. His higher incidence of 91% versus 78% of the patients over 2 kHz in the nimodipine treat-

ed group may result from late surgery. The finding that a greater percentage of patients were in the frequency ranges of between 2 and 4 kHz in the treated patients is consistent with our findings (Fig. 5).

In the dose-control double-blind study we found a less severe vasospasm in all patients and in all subgroups divided according to source of the SAH, and the arteries measured (Figs. 1–4). There is a clear frequency shift of the 3 mg group to 2–4 kHz, while the 2 mg group patients were categorized more in the 4–5.9 kHz group. Astonishing is that the incidence of very severe vasospasm with high frequency ranges between 5 and 7 kHz was not influenced by a higher nimodipine dosage and the incidence of patients with no vasospasm was the same (≤ 2 kHz) (Fig. 5).

The higher number of severe SAH (CT = III) in the 3 mg group may result in a higher incidence of vasospasm, whereas the higher percentage of older patients in the 3 mg group may result in less vasospasm. Our data also reflect the influence of prophylactic hypertension therapy which was induced in 47% of the patients. Only 26% of the patients showed no pathological velocity changes in the period after SAH, which indicates that there is no vasospasm, while the others 74% had vasospasm. We conclude from our findings, as did Seiler and Pasqualin in the discussed studies, that nimodipine or a higher nimodipine dosage reduces the severity of vasospasm, but that it does not influence the incidence of vasospasm. Prophylactic nimodipine treatment results therefore in a better clinical outcome, as has been shown in numerous clinical studies[4, 8, 9, 15, 16, 33, 34].

Is the reduction of the velocities caused by a vessel lumen increase of the large arteries at the base of the skull, or does it reflect a dilating effect of the "small resistance" small vessels? If the resistance vessels were dilated, this would make us suspect not a decrease, but rather an increase in velocity in the MCA. If this opening of collaterals were the cause of the velocity reduction, we would expect to find an increase in A 1 or P 1 on the same or contralateral hemisphere. However, this did not occur[21].

As was shown in a previous study[20] changing the application of nimodipine from intravenous to oral causes a secondary velocity increase to occur after the intravenous administration has been discontinued. This indicates that the effect of the drug has been lowered, resulting in a vessel lumen narrowing again. We think that our findings can be explained by the fact that the large arteries of the circle of Willis dilate slightly, which cannot be detected in angiographic studies[4, 27, 33]. But the haemodynamic sign of a slight dilation of the large arteries caused by nimodipine results in an increase in volume flow and a reduction of the flow velocity.

References

1. Aaslid R, Markwalder Th-M, Nornes H (1982) Noninvasive transcranial Doppler ultrasound recording of flow velocity in basal cerebral arteries. J Neurosurg 57: 769–774
2. Aaslid R, Huber P, Nornes H (1984) Evaluation of cerebrovascular spasm with transcranial Doppler ultrasound. J Neurosurg 60: 37–41
3. Aaslid R (1986) Transcranial Doppler sonography. Springer, Wien New York
4. Allen GS, Hyo SA, Preziosi TJ, Battye R, Boone SC, Chou SN, Kelly DL, Weir BK, Crabbe RA, Lavik PJ, Rosenbloom SB, Dorsey FC, Ingram CR, Mellits DE, Bertsch LA, Boisvert PJ, Hundley MB, Johnson RK, Strom JA, Transou CR (1983) Cerebral arterial spasm – a controlled trial of nimodipine in patients with subarachnoid haemorrhage. N Engl J Med 308: 619–624
5. Auer LM (1981) Pial arterial vasodilation by intravenous nimodipine in cats. Drug Res 31: 1423–1425
6. Auer LM, Oberbauer RW, Schalk HV (1983) Human pial vascular reactions to intravenous nimodipine-infusion during EC-IC-bypass surgery. Stroke 14: 210–213
7. Auer LM, Suzuki A, Yasui N, Ito Z (1984) Intraoperative topical nimodipine after aneurysm clipping. Neurochirurgia 27: 36–38
8. Auer LM (1984) Acute operation and preventive nimodipine improve outcome in patients with ruptured cerebral aneurysms. Neurosurgery 15: 57–66
9. Auer LM, Brandt L, Ebeling U, Gilsbach J, Groeger H, Harders A, Ljunggren B, Oppel F, Reulen HJ, Säveland H (1986) Nimodipine and early aneurysm operation for good condition SAH patients. Acta Neurochir (Wien) 82: 7–13
10. Ecker A, Riemenschneider PA (1951) Arteriographic demonstration of spasm of the intradural arteries with special reference to saccular arterial aneurysms. J Neurosurg 8: 660–667
11. Fisher CM, Roberson GH, Ojemann RG (1977) Cerebral vasospasm with ruptured saccular aneurysm – the clinical manifestations. Neurosurgery 1: 245–248
12. Fisher CM, Kistler JP, Davis JM (1980) Relation of cerebral vasospasm to subarachnoid haemorrhage visualized by computerized tomographic scanning. Neurosurgery 6: 1–9
13. Fujisawa A, Matsumoto M, Matsuyama T, Ueda H, Wanaka A, Yoneda S, Kimura K, Kamada T (1986) The effect of the calcium antagonist nimodipine on the gerbil model of experimental cerebral ischaemia. Stroke 17: 748–752
14. Gilsbach J, Harders A, Hornyak ME (1987) Cerebral vascular spasm in aneurysm surgery and its clinical significance. In: Sinha KK, Chandra P (eds) Progress in clinical neurosciences, vol 1. Ranchi, Catholic Press, pp 125–133
15. Gilsbach JM, Harders A, Eggert HR (1988) Early aneurysm surgery: a 7 year clinical practice report. Acta Neurochir (Wien) 90: 91–102
16. Gilsbach JM, Ljunggren B, v Holst H, Seiler R, Mokry M, v Essen C, Conzen M (1987) Outcome in 204 SAH patients subjected to early aneurysm surgery and intravenous nimodipine. Report on a multicenter double blind dose comparison study. Presented at the 8th European Congress of Neurosurgery, Barcelona, Spain, September 6–11
17. Gilsbach J, Harders A, Hornyak M (1988) Does vasospasm cause major morbidity and mortality in early aneurysm surgery? In: Wilkins RH (ed) Proceedings on cerebral vasospasm. Raven Press, New York (in press)
18. Gilsbach J (1988) Nimodipine in the prevention of ischaemic deficits after aneurysmal subarachnoid haemorrhage. Acta Neurochir (Wien) [Suppl] 45: 41–50
19. Harders A (1986) Neurosurgical applications of transcranial Doppler sonography. Springer, Wien New York
20. Harders A (1986) Erfassung der Wirkung von Nimodipin bei Patienten mit Subarachnoidalblutung und Aneurysma-Frühoperationen mittels transkranieller Dopplersonographie. In: Kazner E (ed) Nimotop in der Prophylaxe und Therapie neurologischer Ausfälle durch zerebrale Ischämie, vol 4. G Braun, Karlsruhe, pp 105–112
21. Harders A, Gilsbach J (1987) Time course of blood velocity changes related to vasospasm in the circle of Willis measured with the noninvasive transcranial Doppler method. J Neurosurg 66: 718–728
22. Harders A, Gilsbach J, Hornyak M (1988) Incidence of vasospasm in transcranial Doppler sonography and its clinical significance. In: Wilkins RH (ed) Proceedings on cerebral vasospasm. Raven Press, New York (in press)
23. Harper AM, Craigen L, Kazda S (1981) Effect of the calcium antagonist, nimodipine, on cerebral blood flow and metabolism in the primate. J Cereb Blood Flow Metabol 1: 349–356
24. Hunt WE, Hess RM (1968) Surgical risk as related to time of intervention in the repair of intracranial aneurysms. J Neurosurg 28: 14–20
25. Kodama N, Mizoi K, Sakurai Y, Suzuki J (1980) Incidence and onset of vasospasm. In: Wilkins RH (ed) Cerebral arterial spasm. Williams and Wilkins, Baltimore London, pp 361–365
26. Lindegaard KF, Nornes H, Bakke SJ, Sorteberg W, Nakstad P (1988) Using transcranial Doppler recordings in the evaluation of patients at risk for cerebral vasospasm after subarachnoid haemorrhage. 2nd International Symposium on Intracranial Haemodynamics, San Diego, February 16–18
27. Ljunggren B, Brandt L, Säveland H, Nilsson PE, Cronqvist S, Andersson KE, Vinge E (1984) Outcome in 60 consecutive patients treated with early aneurysm operation and intravenous nimodipine. J Neurosurg 61: 864–873
28. Ljunggren B, Brandt L, Säveland H (1988) Nimodipine in aneurysmal subarachnoid haemorrhage. Cerebral vasospasm – 1987: a research conference; Charlottesville, Congress book, accepted
29. Meyer FB, Anderson RE, Yaksh TL, Sundt TM (1986) Effect of nimodipine on intracellular brain pH, cortical blood flow, and EEG in experimental focal cerebral ischaemia. J Neurosurg 64: 617–626
30. Mohamed AA, Mendelow AD, Teasdale GM, Harper AM, McCulloch J (1985) Effect of the calcium antagonist nimodipine on local cerebral blood flow and metabolic coupling. J Cereb Blood Flow Metabol 5: 26–33
31. Müller-Schweinitzer E, Neumann P (1983) Effects of calcium antagonists PN 200–210, nifedipine, and nimodipine on human and canine cerebral arteries. J Cereb Blood Flow Metabol 3: 354–361
32. Pasqualin A, Barone G, Battaglia R, Da Pian R (1987) Clinical trial on the anti-ischaemic effects of nimodipine in SAH patients.

Presented at the European Course in Neurosurgery, Satellite Symposium, Wroclaw, Poland, June 27–July 3, 1987

33. Philippon J, Grob R, Dagreou F, Guggiari M, Rivierez M, Viars P (1986) Prevention of vasospasm in subarachnoid haemorrhage. A controlled study with nimodipine. Acta Neurochir (Wien) 82: 110–114

34. Säveland H, Ljunggren B, Brandt L, Messeter K (1986) Delayed ischaemic deterioration in patients with early aneurysm operation and intravenous nimodipine. Neurosurgery 18: 146–150

35. Saito J, Sano K (1980) Vasospasm after aneurysm rupture: Incidence, onset and course. In: Wilkins RH (ed) Cerebral arterial spasm. Williams and Wilkins, Baltimore London, pp 294–301

36. Seiler RW, Grolimund P, Aaslid R, Huber P, Nornes H (1986) Cerebral vasospasm evaluated by transcranial ultrasound correlated with clinical grade and CT visualized subarachnoid haemorrhage. J Neurosurg 64: 594–600

37. Seiler RW, Grolimund P, Zurbruegg H (1987) Evaluation of the calcium-antagonist nimodipine for the prevention of vasospasm after aneurysmal subarachnoid haemorrhage. A prospective transcranial Doppler ultrasound study. Acta Neurochir (Wien) 83: 8–16

38. Seiler RW, Grolimund P, Weber M (1988) Effect of nimodipine on the CO_2-reactivity of patients with aneurysmatic subarachnoid haemorrhage: A transcranial Doppler study. 2nd International Symposium on Intracranial Haemodynamics. San Diego, February 16–18

39. Weir B (1980) The incidence and onset of vasospasm after subarachnoid haemorrhage from ruptured aneurysm. In: Wilkins RH (ed) Cerebral arterial spasm. Williams and Wilkins, Baltimore London, pp 302–305

Correspondence: Dr. A. Harders, Neurochirurgische Universitätsklinik, D-7800 Freiburg i. Br., Federal Republic of Germany.

Acta Neurochirurgica, Suppl. 45, 29–35 (1988)
© by Springer-Verlag 1988

Interactions Between Nimodipine and General Anaesthesia – Clinical Investigations in 124 Patients During Neurosurgical Operations

H. Müller[1], **H. Kafurke**[1], **P. Marck**[1], **J. Zierski**[2], and **G. Hempelmann**[1]

[1] Department of Anaesthesiology and Intensive Care Medicine and [2] Department of Neurosurgery, Justus-Liebig-University, Giessen, Federal Republic of Germany

Summary

Haemodynamic, respiratory, metabolic and endocrine investigations were performed in a total number of 124 patients, divided into four different groups, during opiate anaesthesia for neurosurgical operations in order to characterize general effects of nimodipine, a calcium channel blocking agent with a preferential cerebrovascular action. These studies led to the following conclusions: Nimodipine is a vasodilator drug with a hypotensive action, which is especially obvious in hypertensive patients and in combination with similarly acting agents, such as sodium nitroprusside or nitroglycerine. This vascular hypotensive effect may be also enhanced by combined cardiodepressive activity if nimodipine is applied together with inhaled anaesthetics, such as halothane or isoflurane. Nimodipine as well as other vasodilator drugs may lead to increased pulmonary shunting in patients with artificial ventilation, which, however, can be reduced by adequate positive endexspiratory pressure. With high doses the decrease of oxygen extraction and consumption, seen with nimodipine, is accompanied by a moderate rise of lactate. Determination of stress hormones did not reveal analgesia potentiation of nimodipine, as has been assumed in other studies.

Keywords: Calcium antagonist; nimodipine; general anaesthesia; opiate anaesthesia; neuroanaesthesia.

Introduction

Most studies on nimodipine, including those presented in this journal, are primarily dealing with the particular, predilective effects of this calcium antagonist upon cerebral vessels and tissue and their clinical implications. Other, more general actions of this drug are often looked upon as a kind of side-effects, although they represent important aspects of the basic calcium antagonistic pharmacology of nimodipine.

We have investigated the haemodynamic, respiratory, metabolic and endocrine activity of nimodipine during general anaesthesia for neurosurgical operations with regard to the following questions:

– What is the role of nimodipine, either alone or together with vasodilative drugs (sodium nitroprusside or nitroglycerine), for intraoperative hypotension in normotensive or hypertensive patients?

– Are there interactions with opiates or inhaled anaesthetics, such as halothane or isoflurane?

– Does nimodipine influence pulmonary circulation or gas exchange?

– Are there metabolic alterations either due to the lowering of blood pressure or as a consequence of a direct action of this drug?

– Are there endocrine interactions, *e.g.* in analgesia?

– What role has alcohol, the solvent of the preparation, for general activities of nimodipine?

Methods and Patients

To answer these questions we studied a total number of 124 patients. All of them had given their informed consent and received intravenous nimodipine in doses between 0.4 and 1.2 µg/kg/min during standardized general anaesthesia (introduction with thiopentone, relaxation with pancuronium bromide, analgesia with fetanyl 0.01 mg/kg initially and 0.0015 mg/kg every 45 min, artificial ventilation with oxygen and nitrous oxide) for neurosurgical operations (trepanation). Since the standard dose of nimodipine is 0.5 µg/kg/min the administered dose was up to three times higher. Patients with pre-existing cardiovascular or pulmonary diseases were excluded from the study.

In all patients the following haemodynamic parameters were registered by invasive techniques: SAP (systolic arterial pressure), DAP (diastolic arterial pressure), HR (heart rate), CO (cardiac output), SPAP (systolic pulmonary arterial pressure), DPAP (diastolic pulmonary arterial pressure), PCWP (pulmonary capillary wedge pressure = left atrial pressure), CVP (central venous pressure = right atrial pressure). These values were used for calculation of additional haemodynamic parameters: MAP (mean arterial pressure), CI (cardiac index), SV (stroke volume), SI (stroke index), MPAP (mean pulmonary arterial pressure), SVR (systemic vascular resistance), PVR (pulmonary vascular resistance), LVSWI (left ventricular stroke work index), RVSWI (right ventricular stroke work index), LCWI (left cardiac work index), RCWI (right cardiac work index), RPP (rate pressure product), TI (triple index). Arterial and mixed venous blood gas analyses were used for determination of p_aO_2 (arterial oxygen pressure), p_aCO_2 (arterial carbon dioxide pressure), p_vO_2 (mixed venous oxygen pressure) at a given haemoglobin (Hb) and fraction of inspired oxygen (FiO_2). Additional respiratory parameters were calculated from these data: $AVDO_2$ (arterial-venous oxygen difference), VO_2 (oxygen consumption), Q_S/Q_T (pulmonary shunting), C_aO_2 (arterial oxygen concentration), C_vO_2 (venous oxygen concentration), O_2ER (oxygen extraction), O_2AV (oxygen availability). Blood samples were taken for determination of glucose, lactate, free glycerol, albumine, antidiuretic hormone (ADH), adrenocorticotropic hormone (ACTH) and cortisol.

Controlled Hypotension by Nimodipine and Sodium Nitroprusside

In the first group studies were done during clipping operations in patients with cerebral aneurysms. Controlled hypotension during opiate anaesthesia and ventilation without positive endoexspiratory pressure (PEEP) was either done with sodium nitroprusside (SNP) alone (maximal dosage: 8 µg/kg/min, group 1 A, n = 11) or with a combination of nimodipine and SNP (group 1 B, n = 11). In the latter patients nimodipine was given first and up to a maximal dosage of 1.2 µg/kg/min, while SNP was added later on and in an amount necessary to achieve a definite reduction of blood pressure. The different measuring points were not related to time but to a definite reduction of systolic blood pressure (I = initial value after start of operation and without hypotensive medication, II resp. III = systolic pressure at 100 resp. 90 mmHg, IV–V = systolic blood pressure of 80 mmHg in an interval of 20 min, VI = value before the end of operation).

Effects in Normo- and Hypertensive Patients

In the second group we investigated the efficacy of nimodipine for treatment for intraoperative hypertension during long-term neurosurgical procedures in standardized opiate anaesthesia without PEEP-ventilation. Nimodipine was infused in normotensive (group 2 A, n = 17) and hypertensive patients (group 2 B, n = 14). Arterial hypertension was known from the clinical history and had been pretreated with antihypertensive drugs and/or beta blockers. An additional group of 11 patients (group 2 C) received an infusion of placebo containing alcohol as the solvent of the preparation. Measuring was done at intervals of 30 min, first without nimodipine or placebo (I), then 30 min after 0.4 µg/kg/min of nimodipine (or the same volume of placebo) (II) and another 30 min after 0.7 µg/kg/min of nimodipine (III). After that nimodipine was stopped (IV = 30 min after the end of nimodipine infusion).

Comparison Nimodipine with Nitroglycerine

In our third group of patients the hypotensive effect of nimodipine (1.0 µg/kg/min, group 3 A, n = 10) was compared with that of nitroglycerine (2.4 µg/kg/min, group 3 B, n = 10). These examinations were also done during opiate anaesthesia for longterm neurosurgical operations, but this time PEEP of 5 cm H_2O was provided. An additional control group (group 3 C, n = 10) was included, which received none of the hypotensive drugs. Nimodipine or nitroglycerine were given for a period of up to 3 hours (I = after start of operation without hypotensive medication, II–VI = during continuous infusion of nimodipine or nitroglycerine in intervals of 30 min, VII = at the end of operation without hypotensive medication and during ventilation with oxygen alone (FiO_2 = 1.0)).

Nimodipine and Inhalation Anaesthetics

In our fourth group of patients nimodipine was combined with an inhaled anaesthetic given in addition to standardized opiate anaesthesia with PEEP-ventilation (5 cm H_2O). The inhaled anaesthetics halothane (group 4 A, n = 10) or isoflurane (group 4 B, n = 10) were given in equivalent concentrations (MAC 1). A control group (group 4 C, n = 10) did not receive halothane or isoflurane. The inhaled anaesthetics were started 30 min before the first measuring point (I = without nimodipine), while measuring II and III were done at the end of an 30 min infusion cycle with nimodipine 0.4 and 0.7 µg/kg/min. After that nimodipine was stopped (IV = 30 min after the end of nimodipine infusion but with continuing application of inhaled anaesthetics).

Results

Controlled Hypotension by Nimodipine and Sodium Nitroprusside

Some of the results of the first group (comparison of SNP and SNP-nimodipine for controlled hypotension)

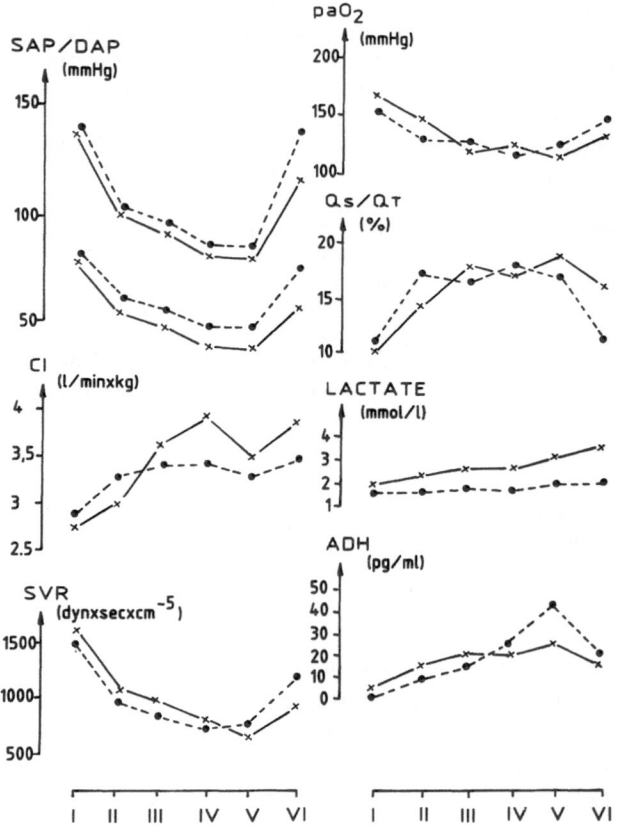

Fig. 1. Group 1 (sodium nitroprusside alone: group 1 A: n = 11: ● – – – ● – – – ● / sodium nitroprusside and nimodipine: group 1 B: n = 11: × —— × —— ×): systolic arterial pressure (SAP), diastolic arterial pressure (DAP), cardiac index (CI), systemic vascular resistance (SVR), arterial oxygen pressure (p_aO_2), pulmonary shunting (Q_S/Q_T), lactate and antidiuretic hormone (ADH). Explanation of method and measuring points I–VI: see text

Table 1. *Average Dosage of Vasodilator Drugs for Controlled Hypotension in Group 1* = (Methods and measuring points I–VI are in detail explained in the text)

	I	II	III	IV	V	VI
Group 1 A (SNP alone)						
SNP (μg/kg/min)	0	3.30	4.88	6.23	6.38	0
Group 1 B (SNP + nimodipine)						
SNP (μg/kg/min)	0	0.62	1.09	1.84	1.71	0
Nimodipine (μg/kg/min)	0	0.95	1.10	1.16	1.19	0.70
	I	II	III	IV	V	VI
	initial value	p_{syst} = 100 mmHg	p_{syst} = 90 mmHg	p_{syst} = 80 mmHg	p_{syst} = 80 mmHg	value before end of operation

are summed up in Fig. 1. As the different measuring points were related to a definite reduction of blood pressure (maximal value of controlled hypotension: 80 mmHg systolic blood pressure), there of course was no difference between both groups with regard to systemic blood pressure. In both groups, but much more pronounced in group 1 A (SNP alone), a consecutive increase in heart rate occurred during hypotension. Pulmonary arterial pressure and right and left atrial pressure were only slightly decreased in both groups. There was also no difference with regard to cardiac and stroke indices and volumes, which were increased in both groups. Systemic vascular resistance decreased significantly, while pulmonary vascular resistance remained almost stable. Cardiac work indices as well as parameters of contractility (RPP, TI) were reduced in both groups, demonstrating improved conditions for cardiac contraction with less energy consumption. There was a decrease in P_aO_2 in both groups, while p_aCO_2 and p_vO_2 remained stable. Parallel to the reduction of arterial oxygen pressure during hypotension, oxygen concentration and $AVDO_2$ decreased. At the same time oxygen consumption and extraction were reduced and oxygen availability was increased (the latter changes were more pronounced in group 1 B, in which nimodipine was combined with SNP). The decrease in arterial pO_2 may be referred to the increase in pulmonary shunting, which was independent from the used vasodilator drug. Free glycerol was only increased by SNP and lactate showed a constant rise in the SNP/nimodipine group. Glucose was similarly increased in both groups during hypotension. There were no differences with regard to the hormones ADH, ACTH and cortisol, which were increased in both groups. Albumine and haemoglobin showed a slow decrease during the course of the operation, demonstrating that the observed changes of hormones and metabolic parameters were independent from dilution and intraoperative blood loss. The overall statistical comparison of both groups

(variance analysis) demonstrated significant differences only with regard to free glycerol, lactate and diastolic arterial blood pressure (which was more reduced in group 1 B). The most important aspect of this com-

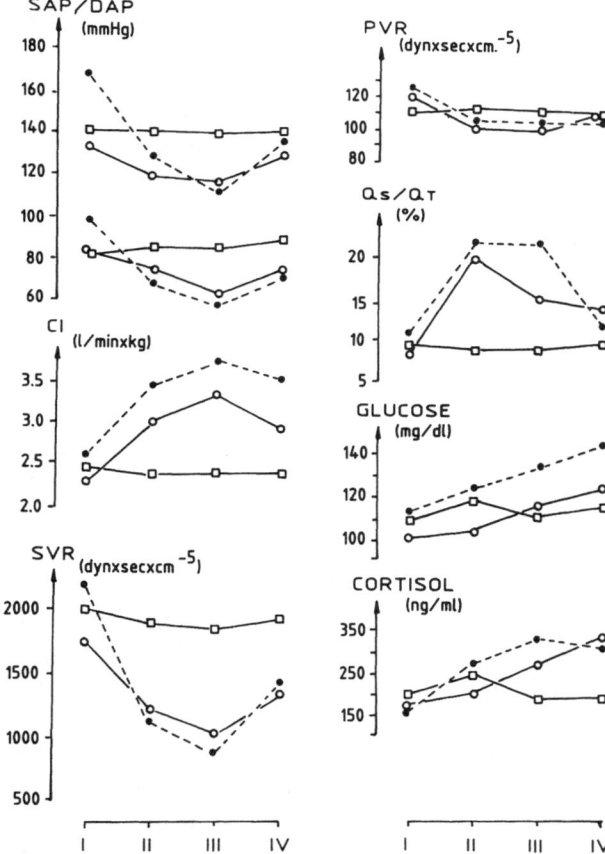

Fig. 2. Group 2 (normotensive patients: group 2 A: n = 17: ○———○———○ / hypertensive patients: group 2 B: n = 14: ●– – –●– – –● / placebo: group 2 C: n = 11: □———□———□): systolic arterial pressure (SAP), diastolic arterial pressure (DAP), cardiac index (CI), systemic vascular resistance (SVR), pulmonary vascular resistance (PVR), pulmonary shunting (Q_S/Q_T), glucose and cortisol. Explanation of method and measuring points I–IV: see text

parison seem to be the reduction of SNP dosage by 70–80% due to the addition of the vasodilative calcium-antagonist nimodipine (Table 1).

Effects in Normo- and Hypertensive Patients

In the second group of patients (Fig. 2) the hypotensive effect of nimodipine was compared in normo- and hypertensive patients. As some of the results are similar to that in group 1, not all parameters will be described for group 2. There were no changes in any of the parameters with placebo, *i.e.* alcohol. Nimodipine reduced blood pressure especially in hypertensive patients, *i.e.* a similarly reduced blood pressure was obtained in both groups (2 A and 2 B) during nimodipine, although the initial values before infusion of the calcium antagonist were completely different. Stroke and cardiac index were increased by a decrease in cardiac afterload, as demonstrated by a reduction of vascular resistance (which started from a much higher level in the group with hypertensive patients). A slight decrease in pulmonary vascular resistance was also present. The respiratory changes after nimodipine in normo- and hypertensive patients were similar to that in group 1 and pulmonary shunting seemed to be more pronounced in hypertensive patients. There also was an increase in lactate and blood glucose, but no changes in free glycerol. ADH, ACTH and cortisol were similarly increased as in group 1.

Comparison Nimodipine with Nitroglycerine

In group 3 (Fig. 3) the hypotensive effect of nimodipine was compared with nitroglycerine in an equipotent dosage. The control group reflected the usual changes during opiate anaesthesia without additional medication. Nimodipine and nitroglycerine caused a similar reduction of blood pressure with an initial increase in heart rate, which was more pronounced with nitroglycerine. While nimodipine enhanced cardiac and stroke indices, these parameters were reduced by nitroglycerine. Systemic vascular resistance was diminished with nimodipine, but increased with nitroglycerine. On the other hand, nitroglycerine caused a more pronounced decrease of pulmonary resistance. In contrast to the already described groups p_aO_2 remained more or less stable. The increase at the end of the operation is due to ventilation with oxygen alone ($FiO_2 = 1.0$). In addition, the enhancement of pulmonary shunting was not present in group 3. The increase of shunting at the end of the operation, *i.e.* during pure oxygen ventilation, had nothing to do with

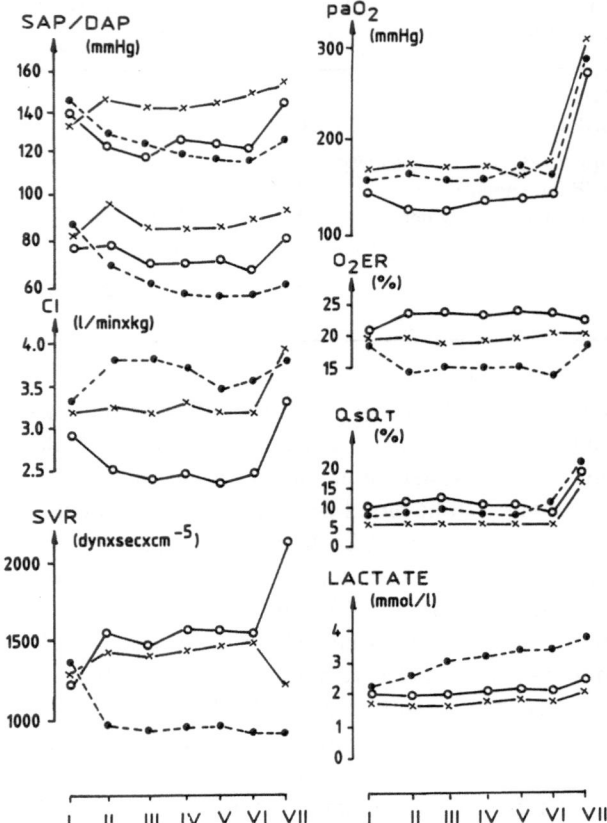

Fig. 3. Group 3 (nimodipine: group 3 A: n = 10: ●– – –●– – –● / nitroglycerine: group 3 B: n = 10: ○——○——○ / control group: group 3 C: n = 10: ×——×——×): systolic arterial pressure (SAP), diastolic arterial pressure (DAP), cardiac index (CI), systemic vascular resistance (SVR), arterial oxygen pressure (p_aO_2), oxygen extraction (O_2ER), pulmonary shunting (Q_S/Q_T) and lactate. Explanation of method and measuring points I–VII: see text

actions of the drugs and could also be demonstrated in the control group. Similar to group 1 and 2, significant reduction of oxygen extraction occured with nimodipine (group 3 A), leading to an increase of oxygen availability. In addition, there was a slow, but steady rise in the anaerobic metabolic derivative lactate, which became especially obvious in this group by long-term measurement for more than 3 hours. This increase was not present with nitroglycerine or in the control group (groups 3 B and 3 C).

Nimodipine and Inhalation Anaesthetics

In the fourth group of patients (Fig. 4) nimodipine was combined with inhaled anaesthetics. A maximum drop in blood pressure occurred with halothane, but also with isoflurane the lowering of blood pressure was

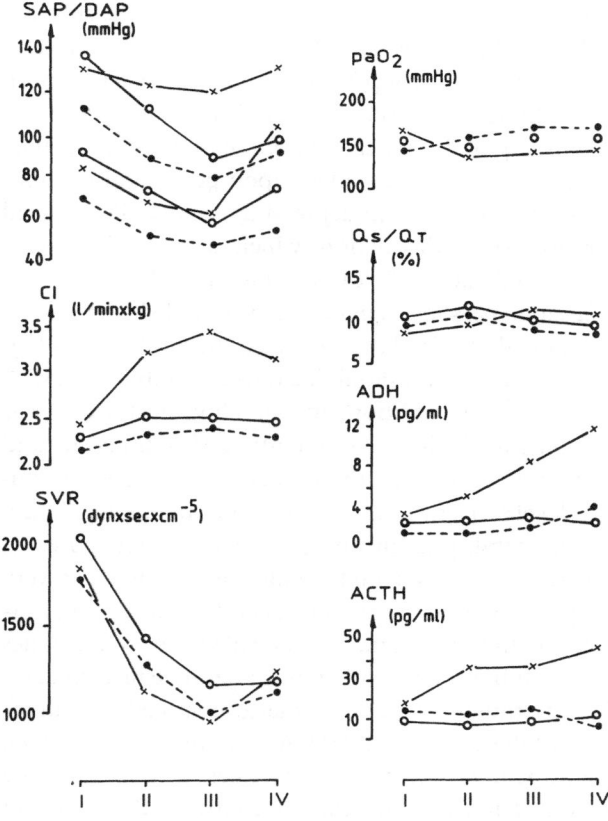

Fig. 4. Group 4 (halothane: group 4 A: n = 10: ○———○———○ / isoflurane: group 4 B: n = 10: ●− − −●− − −● / control group: group 4 C: n = 10: ×———×———×): systolic arterial pressure (SAP), diastolic arterial pressure (DAP), cardiac index (CI), systemic vascular resistance (SVR), arterial oxygen pressure (p_aO_2), pulmonary shunting (Q_S/Q_T), antidiuretic hormone (ADH) and adrenocorticotropic hormone (ACTH). Explanation of method and measuring points I–IV: see text

Discussion

Our studies demonstrate, that nimodipine may be safely used for intraoperative reduction of an elevated blood pressure and to reduce the requirement of other vasodilator agents in different situations during anaesthesia, including controlled hypotension for surgery of cerebral aneurysm. In this regard our results confirm previous results of other authors[14]. In addition, reflex tachycardia, which may occur with SNP or nitroglycerine, are not to be expected with nimodipine, which in addition to its peripheral vascular action has some mild cardiac depressive actions. Arterial vasodilation, however, seems to be the major effect of this drug, responsible for the haemodynamic actions. Hypotensive action may be demonstrated especially in case of pre-existing vasoconstriction, *e.g.* in hypertensive patients[4]. This particular effect of nimodipine also becomes obvious during combination with other vasodilator drugs, such as SNP, which although being a potentially toxic drug (it produces cyanide[15]) is the most often used agent for controlled hypotension[10]. The combination of SNP with nimodipine may help to reduce the dosage of SNP without introducing disadvantages with regard to haemodynamic, respiratory, metabolic and endocrine reactions. The increase in free glycerol, observed in group 1 A (SNP alone) may be an example of a metabolic inhibitory interaction of SNP.

The observed haemodynamic effects of nimodipine, as well as of SNP and nitroglycerine, need further explanation: Arterial vasodilation, as induced by nimodipine and SNP, leads to a decrease in systemic vascular resistance and cardiac afterload and therefore to an increase in cardiac output, while the cardiac work indices are typically reduced. On the other hand, nitroglycerine is known to act by venous vasodilation, leading to venous pooling, which will reduce cardiac preload and decrease cardiac output. With both mechanisms cardiac work and contractility are reduced or, as it is often said, used more economically. Venous pooling will also lead to a more pronounced decrease of pulmonary vascular resistance. Of course, arterial or venous vasodilation will both induce reduction of blood pressure, *i.e.* hypotension. High concentrations of alcohol may also be responsible for arterial vasodilation. We were, however, not able to demonstrate relevant haemodynamic actions of the placebo-solution of nimodipin, which contains alcohol.

Calcium-antagonists may have cardiac depressant effects. Nevertheless, in the case of nimodipine such

much more pronounced than in the control group, which received nimodipine but none of the inhaled anaesthetics. In addition, the increase of cardiac index, which we saw in the other groups after nimodipine and which was also present in the control group (group 4 C), was almost not demonstrable during the combination of nimodipine and inhaled anaesthetics. There was, however, no difference to control with regard to the reduction of systemic vascular resistance. Pulmonary shunting and arterial pO_2 were not changed by nimodipine in group 4, but the alterations of O_2ER and O_2AV, which had been typical for nimodipine in the other groups, were still present. The increase in ADH, ACTH and cortisol, which also had been characteristic in the other groups, was prevented by the addition of inhaled anaesthetics to the opiate anaesthesia during the infusion of nimodipine.

effects are not obvious during types of anaesthesia, which are free from cardiac influences, such as opiate anaesthesia, used in our studies as basic anaesthesia in all patients. Combination with inhaled anaesthetics, which are known to impair cardiac calcium influx and therefore are negative inotropic, may unmask the cardiac depressive effect of nimodipine. Such combined depressive effects have been observed for different calcium antagonists in animal experiments[9, 11] and are of particular interest, because they may allow one to characterize more precisely the cardiac and cerebral activities of volatile anaesthetics. In our clinical studies we had a marked drop of blood pressure during the combined use of nimodipine and halothane or isoflurane. At the same time, the increase of cardiac index, usually present during nimodipine infusion, could no longer be observed.

Pharmacological vasodilation will usually be present in the systemic and pulmonary circulation (although the latter effect will sometimes not be evident by registration of pulmonary vascular resistance). Therefore, the use of vasodilator or antihypertensive drugs will lead to increased pulmonary shunting and, as a consequence of this, to a decrease in arterial oxygen pressure. Usually these changes are minimal, but they may become relevant during artificial respiration, independent from the used drug but in close relation to the overall hypotensive effect[3, 12, 16]. Pulmonary shunting has also been observed during continuous infusion of nimodipine in ventilated patients[8], but not in spontaneous breathing patients[6]. In our studies shunting was evident in patients without PEEP-ventilation, but could be almost completely prevented by PEEP (group 3 and 4). PEEP is known to reduce the formation of micro-atelectases and produces an adequate relation between ventilation and circulation of the lungs. Therefore it may be concluded, that pulmonary vasodilation, as seen after all systemic vasodilator drugs including calcium antagonists, will increase pulmonary shunting, if micro-atelectases are present and they are present during ventilation without PEEP. Therefore it seem to be useful to provide PEEP if vasodilation is necessary in a ventilated patient.

Already several years ago it had been demonstrated that calcium antagonists reduce oxygen consumption and mitochondrial oxidative phosphorylation at least in muscle-contraction, but may be also during and after brain ischaemia[5, 13]. In all our groups and probably independent from the degree of hypotension, nimodipine induced a decrease in oxygen extraction and an increase in oxygen availability, while at the same time

a slow rise in lactate could be observed. The increase in lactate became especially obvious during long term measurement in group 3 and was neither seen with SNP nor with nitroglycerine. Therefore the question arises, if nimodipine by its calcium antagonistic action may have an unfavourable effect upon peripheral metabolism, *i.e.* a metabolism depressing activity. The clinical importance of this action, which seems to be absent during infusion of lower, standard doses and in conscious persons, remains to be established.

Animal studies have reported on analgesia potentiating actions of nimodipine during combination with opioids[1, 7]. In addition, in one clinical study it was demonstrated, that a combination of nimodipine and low-dose opiate anaesthesia was accompanied by similar levels of stress hormones and similar haemodynamic stability as in anaesthesia with high doses of opiates alone[2]. Haemodynamic stability may be very simply explained by the peripheral cardiovascular effects of this drug and the so-called stress hormones may be influenced by other intraoperative mechanisms. In our studies we were unable to demonstrate alterations of the hormones ADH, ACTH and cortisol, which might be interpreted as a consequence of an analgesic action of nimodipine. On the contrary, there was an increase of these hormones, especially ADH, which possibly may be explained as an regulatory effect upon hypotension. This increase was only reduced, if the basis opiate anaesthesia was combined and therefore deepened by addition of inhaled anaesthetics.

In summary, our studies have demonstrated, that nimodipine besides its cerebrovascular effects is characterized by typical qualities of calcium antagonists in general. The relatively high doses given may have helped to reveal some effects which usually are only demonstrable to a moderate degree. In addition, the combination of nimodipine with anaesthesia has possibly led to more pronounced actions of the calcium antagonist. Nevertheless, it may be argued, that under certain clinical conditions and in some patients with pre-existing cardiovascular or respiratory diseases these general effects of nimodipine could become evident and may be the reason for relevant side-effects even during moderate doses of this drug. Parenteral use of nimodipine in higher doses requires adequate haemodynamic and respiratory supervision. This, of course, is also necessary during infusion of other vasodilator agents, such as SNP or nitroglycerine. Induced hypotension during operations should be restricted to those situations, in which the usefulness of this regimen has been demonstrated without a doubt.

References

1. Benedek G, Sziksai M (1984) Potentiation of thermoregulation and analgesic effects of morphine by calcium antagonists. Pharmacol Res Commun 16: 1009–1018
2. Bormann B von, Boldt J, Sturm G, Kling D, Weidler B, Lohmann E, Hempelmann G (1985) Calcium-Antagonisten in der Anaesthesie − Additive Analgesie durch Nimodipin während cardiochirurgischer Eingriffe. Anaesthesist 34: 429 − 434
3. Casthely PA, Lear S, Cottrell JE, Lear E (1982) Intrapulmonary shunting during induced hypotension. Anaesth Analg 61: 231–235
4. Engberding R, Bender F, Gülker H, Specker E, Molinski M (1984) Haemodynamic effects of the new calcium channel blocking agent nimodipine in normotensive and hypertensive subjects. Z Kardiol 73: 498–503
5. Fleckenstein A (1983) Calcium antagonism in heart and smooth muscle. Experimental facts and therapeutic prospects. John Wiley & Sons, New York Chichester Bisbane Toronto Singapore
6. Germann P, Aloy A, Zabloudil B, Richling B, Perneczky A, Zimpfer M (1987) Effekte des Calciumantagonisten Nimodipin auf den pulmonalen Gasaustausch beim neurochirurgischen Intensivpatienten. Anaesthesist 36 [Suppl] V 10.5
7. Guerrero-Munoz F, Fearon Z (1982) Opioids/opiates analgetic response modified by calcium. Life Sci 31: 1237–1240
8. Guggenberger H, Kottler B, Heuser D (1986) Cardio-respiratorische Problematik beim Einsatz eines Calciumantagonisten zur Vasospasmusprophylaxe bei Subarachnoidalblutung. Anaesthesist 35: 429 − 432
9. Kapur PA, Bloor BD, Flacke WE, Olewine SK (1984) Comparison of cardiovascular responses to verapamil during enflurane, isoflurane, or halothane anaesthesia in dogs. Anaesthesiology 51: 156–160
10. Lam AM (1984) Induced hypotension. Can Anaesth Soc J 31: 56–62
11. Rogers K, Hysing ES, Merin RG, Taylor A, Hartley C, Celly JE (1986) Cardiovascular effects of and interaction between calcium blocking drugs and anesthetics in chronically instrumented dogs. Anaesthesiology 64: 568–575
12. Stone GJ, Khambatta HJ, Matteo R (1976) Pulmonary shunting during anaesthesia with deliberate hypotension. Anaesthesiology 45: 508–512
13. Steen PA, Gisvold SE, Milde JH, Newberg LA, Scheithauer BE, Lanier WL, Michenfelder JD (1985) Nimodipine improves outcome when given after complete cerebral ischaemia in primates. Anaesthesiology 62: 406–414
14. Stullken EH, Balastrieri FJ, Prough DS, McWhorter JM (1985) The haemodynamic effects of nimodipine in patients anaesthetized for cerebral aneurysm clipping. Anaesthesiology 62: 346–348
15. Tinker JH, Michenfelder JD (1976) Sodium nitroprusside: pharmacology, toxicology and therapeutics. Anaesthesiology 45: 340–350
16. Zadrobilek E, Spiss CK, Redl G, Draxler VH (1985) Nifedipine induced hypotension in man: intrapulmonary shunting during isoflurane halothane anaesthesia. Anaesthesiology 63 [Suppl] A 521

Correspondence: Prof. Dr. H. Müller, Abteilung für Anästhesiologie und Intensivmedizin, Justus-Liebig-Universität, D-6300 Giessen, Federal Republic of Germany.

Acta Neurochirurgica, Suppl. 45, 36–40 (1988)

Attempts at Prevention and Treatment of Delayed Ischaemic Dysfunction in Patients with Subarachnoid Haemorrhage

R. H. Wilkins

Division of Neurosurgery, Duke University Medical Center, Durham, North Carolina, U.S.A.

Summary

Rupture of an intracranial aneurysm is frequently followed by evidence of intracranial arterial narrowing, which often is accompanied by the delayed onset of a related neurological deficit. During the decades since these detrimental phenomena were first recognized, many attempts have been made to prevent or treat them. Current emphasis in prophylaxis and treatment is on: 1. early operation to eliminate the threat of rebleeding and to allow the gentle removal of as much blood as feasible from the basal subarachnoid cisterns, 2. maintenance or elevation of circulating blood volume, and 3. maintenance or elevation of systemic blood pressure. In recent years, there has also been increasing evidence that the administration of the calcium channel blocking agent nimodipine can reduce the incidence of delayed ischaemic neurological deficits in patients with a ruptured aneurysm.

Keywords: Delayed ischaemic neurological dysfunction; cerebral vasospasm; subarachnoid haemorrhage.

Problems with Diagnosis

Studies of the prevention and treatment of the delayed ischaemic neurological deficit that sometimes follows subarachnoid haemorrhage have been hindered by the inaccuracy of the diagnosis, which in turn has been influenced by clinical inconsistencies and technological limitations. This has also been true for its related phenomenon, cerebral vasospasm.

At present, cerebral vasospasm is usually diagnosed at angiography if the luminal diameter of one or more of the intracranial arteries at the base of the brain is smaller than it is expected to be, and this is not thought to be due to another condition such as hypoplasia[20]. In recent years, an increase in blood flow velocity in the intracranial portion of the internal carotid artery or the proximal portions of the middle or anterior cerebral arteries as measured by transcranial Doppler ultrasound techniques, has also been used as evidence of intracranial arterial spasm[16]. However, neither of these techniques provides accurate information about

blood flow through the brain. Computed tomography (CT) has added more precision to the methods used for measuring regional cerebral blood flow and positron emission tomography has done the same for the measurement of regional cerebral metabolism; yet these techniques are not available in many centers, and therefore they are of limited usefulness in studies of the prevention and treatment of delayed cerebral ischaemia after subarachnoid haemorrhage.

In regard to clinical variability, not all patients with spontaneous subarachnoid haemorrhage develop either arteriographic evidence of cerebral vasospasm or a delayed neurological deficit. Furthermore, not all of the patients with arterial narrowing develop cerebral ischaemia or infarction (symptomatic cerebral vasospasm), either because the narrowing is not severe enough to restrict blood flow sufficiently or because there are adequate compensatory mechanisms to maintain cerebral perfusion[4, 7, 9, 20, 23, 25]. Finally, there are many causes of a delayed neurological deficit in a patient with subarachnoid haemorrhage besides ischaemia related to vasospasm, such as rebleeding, hydrocephalus, hyponatremia, and cerebral oedema[13].

Occurrence

Radiographic evidence of intracranial arterial spasm is encountered in several conditions, most of which result in bleeding within the basal subarachnoid cisterns. The most common association is with a ruptured aneurysm. If the arteriograms made during the routine care of patients with a ruptured intracranial aneurysm are reviewed, narrowing of the lumen of one or more of the intracranial arteries can be detected in at least 35% of the cases. It is rare to find such luminal narrowing during the first day after the patient's first subarachnoid haemorrhage. The peak incidence occurs at about one week after the haemorrhage, with about 60% of the patients showing radiographic evidence of intracranial arterial spasm during the second week. Once present, such arteriographic vasospasm usually resolves gradually within 3 weeks. Luminal narrowing affects primarily the arteries within the subarachnoid space, but can extend retrograde into the proximal arteries. It is most marked in the arteries immediately adjacent to the ruptured aneurysm (focal vasospasm), but frequently involves all of the major arteries at the base of the brain (diffuse vasospasm)[4, 7, 9, 18, 20, 23, 25].

Whether such intracranial arterial narrowing causes cerebral ischaemia or infarction depends on its distribution and severity, and also on the many other factors such as blood pressure, intracranial pressure, blood viscosity, and extent of collateral circulation that also affect circulation and the delivery of oxygen and glucose at the cerebral capillary level. The large number of variables involved have resulted in the conflicting results of studies that have tried to correlate the radiographic and neurological aspects of intracranial arterial spasm. Such studies have assessed morbidity and mortality, alterations in cerebral blood flow, and both radiological and pathological evidence of cerebral infarction[20, 21, 23, 25]. Although the results, especially of the early studies, have been contradictory and confusing, it is now recognized that severe intracranial arterial spasm can reduce cerebral perfusion focally or diffusely, leading to cerebral ischaemia or infarction with the expected neurological deficit or death. It has been estimated that about 20 to 30% of patients with a ruptured aneurysm (i.e., approximately half of the patients who develop arteriographic evidence of vasospasm) will develop related neurological symptoms and signs, and of these almost half will die or have a serious residual neurological deficit[4, 7, 9].

In an assessment of the preliminary data from the International Cooperative Study on Timing of Aneurysm Surgery, Kassell and Torner found that 1,272 of the 3,446 patients died or became disabled[10]. Despite the fact that 80% of the patients were admitted on the day of the aneurysm rupture or the next day, and that approximately 80% of the patients were in good condition on admission, the 6 month outcome evaluation showed that 27.3% had died, 2.2% had a vegetative survival, 5.7% were severely disabled, and 8.4% were moderately disabled. The most significant cause of disability and death was cerebral vasospasm, which accounted for 33.5% of the total; in contrast, the direct effect of the aneurysm rupture was responsible for

25.5% and rebleeding for 17.3%. Among the patients with disability or death due to vasospasm, about half were in each category. This and other studies have pointed out the importance of the detrimental effects of symptomatic vasospasm to the outcome of patients with a ruptured aneurysm.

During the past 15 years, it has been recognized that subarachnoid haemorrhage will occasionally cause morphological alterations in the cerebral arteries[5, 14]. Such structural changes, which include intimal thickening that narrows the arterial lumen, seem to be a nonspecific response to arterial injury. However, they do not appear to be the cause of the usual delayed ischaemic dysfunction syndrome because of their later development after subarachnoid haemorrhage, their distal location and focal distribution, and their relatively minor effect on the arterial lumen.

Prediction

It has become apparent during the past decade that the presence and amount of blood in the basal subarachnoid spaces within 4 days after a spontaneous subarachnoid haemorrhage (as shown by CT scanning or at operation) predict the subsequent occurrence and degree of symptomatic cerebral vasospasm. It is also thought that those patients who do not have CT evidence of blood in the basal subarachnoid spaces do not go on to develop symptomatic vasospasm. This is currently the best established predictor of vasospasm[24].

Another valuable predictor of impending trouble, described more recently, is an increase in blood flow velocity (localized acceleration) in the intracranial portion of the internal carotid artery or the proximal segments of the middle cerebral and anterior cerebral arteries, as measured by transcranial Doppler ultrasound measurements. Seiler et al. noted that all of their patients had a pathologically increased flow velocity (more than 80 cm/sec) from the fourth to tenth days after subarachnoid haemorrhage. Maximum flow velocities in the range of 120 to 140 cm/sec did not lead to brain infarction, but values over 200 cm/sec were associated with such a tendency. Of more importance in their patients was a steep early increase in flow velocity values, which was predictive of severe ischaemia and impending infarction[16].

Other ways of predicting symptomatic cerebral vasospasm have been studied, but these either have not been proven to be of value or have not yet been proven to be of value. Such methods include the demonstration of: 1. cisternal or pericisternal enhancement on a contrast-enhanced CT head scan, 2. an unforeseen reduction in cerebral blood flow, 3. a high level of fibrinogen degradation products in the cerebrospinal fluid (above 80 mcg/ml), 4. a decrease in circulating blood volume and in serum sodium concentration, 5. the presence of diffuse disturbances in the electroencephalogram, and 6. an early disturbance of cerebral vasoreactivity, as measured by a decreased or abolished response of mean hemispheric cerebral blood flow to hyperventilation-induced hypocapnia[12, 24].

Prevention and Treatment

Overview

Many ingenous attempts have been made to prevent or treat symptomatic intracranial arterial spasm[17, 19, 22–24]. Some of these have been directed against agents such as blood and blood components that are presumed to be important in the aetiology and pathogenesis of the condition. Other approaches have been used to improve impaired cerebral circulation no matter what its specific cause.

In 1973, 1980, and 1986, I tabulated the many agents and techniques that have been tried to avert or treat cerebral vasospasm[19, 22, 24]. Such studies have assessed the prevention or reversal of: 1. arterial constriction, 2. changes in arterial morphology, 3. reduction in total or regional cerebral blood flow, 4. radiological or pathological evidence of cerebral ischaemia or infarction, or 5. the development of neurological deficits thought to be on the basis of cerebral ischaemia. They have used systems of increasing complexity from arterial strips in a tissue bath to intact animals, and a variety of species from mice to human beings. The investigations involving drugs or biochemical agents have involved numerous dosages and routes of administration. The approaches taken have been grouped into eight categories, although these categories have changed somewhat between 1973 and 1986.

The most obvious approach, that of searching for a drug to dilate cerebral arteries or antagonize their constriction, has generated a larger number and greater variety of investigations than the other seven approaches together. Yet, despite many promising results in animals, this approach has generally been of limited usefulness in human patients until recently.

Early studies assessed the effects of sympathomimetic amines that stimulate beta-adrenergic receptors, alpha-adrenergic blocking agents, anti-adrenergic agents, catecholamine or serotonin depletors, parasympathomimetic agents, postganglionic cholinergic blocking agents, neuromuscular blocking agents, serotonin antagonists, nitrites, local anaesthetics, and 13 other drugs such as papaverine. Later investigations also evaluated the effects of beta-adrenergic blocking agents, dopamine-beta-hydroxylase inhibitors, phosphodiesterase inhibitors, calcium antagonists, nonsteroidal anti-inflammatory drugs, and 26 other agents. Recent studies have expanded the number of prostaglandins (and agents to influence prostaglandins) tested and have added new drugs and new categories of drugs such as free radical scavengers and disulfide bond-reducing agents. The number of reported experiments and clinical trials involving calcium antagonists increased considerably between my reviews of 1980 and 1986.

The experience with 14 calcium antagonists (calcium channel blockers, calcium entry blockers, calcium modulators) was listed in my 1980 and 1986 reviews. The agent that has received the most attention has been nimodipine. It is commonly used clinically in Europe[3, 11, 15] and has been studied in two clinical trials in the United States[1, 2]. In Canada, it has been shown to have beneficial effects in poor grade patients with a ruptured aneurysm[6]. These studies indicate that nimodipine improves the neurological condition of the patients, but has no major effect on arteriographic cerebral vasospasm. In combination with an early operation to secure a ruptured aneurysm and the use of other modern management techniques, the incidence of delayed ischaemic neurological dysfunction has been reduced. For example, Ljunggren and his colleagues now report an incidence of permanent neurological deficit secondary to ischaemia of only 4%[11].

A second approach to cerebral vasospasm has been to search for drugs to improve cerebrovascular rheology and the oxygen delivery to the brain, and to combat ischaemia and swelling of the brain. Evidence has accumulated that haemodilution is useful in the management of cerebral ischaemia, but the value of this technique has not yet been assessed in human patients with a ruptured aneurysm and existing or threatened delayed ischaemic dysfunction. The same is true for the use of perfluorochemical emulsions.

A third approach has been the use of drugs to prevent fibrinolysis and the intracisternal accumulation of potentially spasmogenic fibrin/fibrinogen degradation products. A fourth, and opposite, tactic has been the use of drugs and procedures to remove blood from the basal subarachnoid cisterns or to neutralize its vasospastic effects.

A fifth approach has been the most effective in dealing with the ischaemic effects of cerebral vasospasm. This has involved the use of drugs to increase blood volume, blood pressure, and cardiac output. There is evidence that patients who have sustained the rupture of an intracranial aneurysm may subsequently become hypovolemic, especially if they are hyponatremic or have been kept at bed rest for several days. The restoration or elevation of blood volume by the intravenous administration of blood or a colloidal solution has the potential of preventing or reversing a delayed ischaemic neurological deficit. Maintaining or elevating the systemic blood pressure has this same potential, but hypertension increases the risk of rerupture of the aneurysm if it has not been clipped.

Procedures to interrupt the sympathetic innervation of the cerebral arteries have not proven to be effective in the prevention or treatment of symptomatic vasospasm in human beings. Neither has the administration of gases, at atmospheric or hyperbaric pressures, to dilate cerebral arteries and increase cerebral oxygenation. Among the other miscellaneous approaches that have been taken to preventing or treating vasospasm, an interesting recent development is the physical dilatation of spastic intracranial arteries with an intravascular balloon catheter.

It was obvious at the 1987 research conference on cerebral vasospasm held at Charlottesville, Virginia that investigators are still pursuing many of the drugs and methods listed in my previous reviews. They are also opening new lines of inquiry that involve new agents and new techniques. Gradually, the detrimental phenomenon of delayed ischaemic dysfunction of the brain following subarachnoid haemorrhage is being brought under control.

Prevention

Among the approaches that have been taken to the prevention of symptomatic cerebral vasospasm, many have not lived up to their initial promise. For example, assuming that subarachnoid blood causes delayed constriction of the arteries at the base of the brain, it seems logical that the early removal of blood from the basal subarachnoid cisterns should prevent the later development of cerebral vasospasm. However, removal of most or all of the extensive and widespread subarachnoid blood has proved difficult, and even vigorous clot removal has not led to the abolition of cerebral vasospasm. Yet, an early operation to clip an intracranial aneurysm does permit the subsequent safe usage of blood volume expansion if impending cerebral ischae-

mia is predicted (such as by an abrupt increase in blood flow velocity in the basal arteries). Among the drugs studied so far, the one most likely to prevent the neurological effects of delayed cerebral ischaemia is the calcium antagonist nimodipine.

Thus, the best current approaches to the prophylaxis of symptomatic intracranial arterial spasm in those patients who are identified to be at significant risk are: 1. an early operation to secure the ruptured aneurysm if the overall condition of the patient permits; gentle mechanical removal of subarachnoid blood clots exposed by the operative approach to the extent possible without injuring vessels, cranial nerves or adjacent brain tissue; and perhaps the insertion of a cisternal drain or drains for use during the initial postoperative period; 2. maintenance of the circulating blood volume; and 3. the administration of nimodipine (in the United States, pending approval for such usage).

Treatment

A reasonable plan of treatment of symptomatic cerebral vasospasm is that reported in 1982 by Kassell and his colleagues[8].

"They began by reducing the intracranial pressure if it was above normal. They administered dexamethasone and employed general supportive measures, including the administration of oxygen as required to keep the arterial pO_2 greater than 70 torr."

"They then accomplished intravascular volume expansion with whole blood or packed red blood cells, supplemented with plasma fractionate or albumin. Crystalloid solutions were infused to maintain serum electrolyte levels in the normal range. Sufficient fluids were given to produce a positive fluid balance and elevate the central venous pressure to approximately 10 torr or the pulmonary artery wedge pressure to 18 to 20 torr. If this regimen did not induce hypertension, an agent such as dopamine or dobutamine was given to raise the blood pressure to a level estimated to exceed that required for reversal of the neurological deficit. After the desired effect was achieved, the blood pressure was allowed to fall to a level just high enough to sustain acceptable neurological function. Fludrocortisone (2 mg per day) or desoxycorticosterone acetate (20 mg per day) was administered in divided doses if needed to help maintain the hypervolemia and hypertension for up to 8 days."

"Kassell and his associates found that the induction of the hypervolemic/hypertensive state usually caused a vagal depressor response, which they blocked with atropine (1 mg i.m. every 3 to 4 hours), and a pronounced diuresis, which they counteracted by giving aqueous vasopressin (5 units i.m. as needed to keep the urine output below 200 ml per hour). If pulmonary oedema developed, they administered digitalis."[24]

Conclusions

About half of the patients who develop the delayed intracranial arterial spasm that frequently follows the rupture of an intracranial aneurysm will develop ischaemic neurological dysfunction, and of these patients almost half will die or have a serious residual neurological deficit. Many ingenious attempts have been made to prevent or to treat symptomatic cerebral vasospasm. Most have failed, but progress has been made in the areas of prediction of occurrence, and of maintenance of the cerebral circulation despite the luminal narrowing of the arteries of the circle of Willis. Although their exact mechanism(s) of action in this condition are not yet known, calcium antagonists have been shown to be beneficial in preventing delayed ischaemic neurological dysfunction. Research efforts are continuing in many countries to pursue the most promising lines of inquiry. There is certainly more reason for optimism now than there was in 1973, in 1980, and even in 1986.

References

1. Allen GS, Ahn HS, Preziosi TJ, Battye R, Boone SC, Chou SN, Kelly DL, Weir BK, Crabbe RA, Lavik PJ, Rosenbloom SB, Dorsey FC, Ingram CR, Mellits DE, Bertsch LA, Boisvert DPJ, Hundley MB, Johnson RK, Strom JA, Transou CR (1983) Cerebral arterial spasm – a controlled trial of nimodipine in patients with subarachnoid hemorrhage. N Engl J Med 308: 619–624
2. Allen GS, Battye R (1985) Unpublished observations
3. Auer LM, Brandt L, Ebeling U, Gilsbach J, Groeger U, Harders A, Ljunggren B, Oppel F, Reulen HJ, Saeveland H (1986) Nimodipine and early aneurysm operation in good condition SAH patients. Acta Neurochir (Wien) 82: 7–13
4. Chyatte D, Sundt TM Jr (1984) Cerebral vasospasm after subarachnoid hemorrhage. Mayo Clin Proc 59: 498–505
5. Conway LW, McDonald LW (1972) Structural changes of the intradural arteries following subarachnoid hemorrhage. J Neurosurg 37: 715–723
6. Disney L (1987) Results of a multi-center double-blind, placebo-controlled trial of nimodipine in poor grade aneurysm patients. Presented at the 37th Annual Meeting of the Congress of Neurological Surgeons, Baltimore, Maryland, October 26
7. Heros RC, Zervas NT, Varsos V (1983) Cerebral vasospasm after subarachnoid hemorrhage: an update. Ann Neurol 14: 599–608

8. Kassell NF, Peerless SJ, Durward QJ, Beck DW, Drake CG, Adams HP (1982) Treatment of ischemic deficits from vasospasm with intravascular volume expansion and induced arterial hypertension. Neurosurgery 11: 337–343
9. Kassell NF, Sasaki T, Colohan ART, Nazar G (1985) Cerebral vasospasm following aneurysmal subarachnoid hemorrhage. Stroke 16: 562–572
10. Kassell NF, Torner JC (1984) The international cooperative study of timing of aneurysm surgery – an update. Stroke 15: 566–570
11. Ljunggren B, Säveland H, Brandt L, Romner B, Andersson KE (1987) Nimodipine in aneurysmal subarachnoid haemorrhage. Presented at cerebral vasospasm – 1987: a research conference, Charlottesville, Virginia, May 1
12. Messeter K, Brandt L, Ljunggren B, Svendgaard NA, Algotsson L, Romner B, Ryding E (1987) Prediction and prevention of delayed ischemic dysfunction after aneurysmal subarachnoid hemorrhage and early operation. Neurosurgery 20: 548–553
13. Peerless SJ (1979) Pre- and postoperative management of cerebral aneurysms. Clin Neurosurg 26: 209–231
14. Peerless SJ, Kassell NF, Komatsu K, Hunter IG (1980) Cerebral vasospasm: Acute proliferative vasculopathy? II. Morphology. In: Wilkins RH (ed) Cerebral arterial spasm. Williams & Wilkins, Baltimore, pp 88–96
15. Philippon J, Grob R, Dagreou F, Guggiari M, Rivierez M, Viars P (1986) Prevention of vasospasm in subarachnoid haemorrhage. A controlled study with nimodipine. Acta Neurochir (Wien) 82: 110–114
16. Seiler RW, Grolimund P, Aaslid R, Huber P, Nornes H (1986) Cerebral vasospasm evaluated by transcranial ultrasound correlated with clinical grade and CT visualized subarachnoid hemorrhage. J Neurosurg 64: 594–600
17. Weir B (1987) Aneurysms affecting the nervous system. Williams & Wilkins, Baltimore, pp 505–569
18. Weir B, Grace M, Hansen J, Rothberg C (1978) Time course of vasospasm in man. J Neurosurg 48: 173–178
19. Wilkins RH (1973) Attempts at treatment of intracranial arterial spasm in animals and human beings. Surg Neurol 1: 148–159
20. Wilkins RH (1975) Intracranial vascular spasm in head injuries. In: Vinken PJ, Bruyn GW (eds) Handbook of clinical neurology, vol 23, Injuries of the brain and skull, part I. North-Holland, Amsterdam, pp 163–197
21. Wilkins RH (1977) The role of intracranial arterial spasm in the timing of operations for aneurysm. Clin Neurosurg 24: 185–207
22. Wilkins RH (1980) Attempted prevention or treatment of intracranial arterial spasm: a survey. Neurosurgery 6: 198–210
23. Wilkins RH (1980) Cerebral arterial spasm. Williams & Wilkins, Baltimore
24. Wilkins RH (1986) Attempts at prevention or treatment of intracranial arterial spasm: an update. Neurosurgery 18: 808–825
25. Wilkins RH, Alexander J, Odom GL (1968) Intracranial arterial spasm: a clinical analysis. J Neurosurg 29: 121–134

Correspondence: Robert H. Wilkins, M.D., Division of Neurosurgery, Box 3807, Duke University Medical Center, Durham, NC 27710, U.S.A.

Acta Neurochirurgica, Suppl. 45, 41–50 (1988)

Nimodipine in the Prevention of Ischaemic Deficits After Aneurysmal Subarachnoid Haemorrhage

An Analysis of Recent Clinical Studies

J. M. Gilsbach

Neurochirurgische Universitätsklinik, Freiburg i. Br., Federal Republic of Germany

Summary

An analysis of the recent trials on the effect of nimodipine on the prevention of delayed ischaemic deficits (DID) due to vasospasm provides strong evidence that the drug reduces the incidence and the severity of DID. Placebo-controlled prospective randomized studies with oral administration of nimodipine prove that patients treated with nimodipine suffer severe disability or death due to vasospasm less frequently than those not treated. Non-controlled open prospective trials with early surgery and intravenous and oral nimodipine present the lowest published incidence of DID in aneurysmal subarachnoid haemorrhage.

Keywords: Subarachnoid haemorrhage; vasospasm; delayed ischaemic deficit; Nimodipine.

Introduction

Cerebral vasospasm following subarachnoid haemorrhage was first described angiographically by Ecker und Riemenschneider in 1951[9]. Kågström[24] first detected the delayed onset of vessel narrowing after aneurysm rupture. The relationship between vasospasm and delayed ischaemic deficits (DID) or delayed neurological dysfunctions (DND) was established by Fisher *et al.* in 1977[11]. Since then, numerous drugs have been advocated to treat symptomatic vasospasm (see Wilkins[46–48]), which occur in up to 50% of cases[11, 19, 21, 29, 41]. With the exception of hypertensive-hypervolaemic therapy[7, 10, 20, 30], none of these drugs and treatment regimen proved satisfactory. Early aneurysm surgery, introduced in the late 1970's, reduced the risk of re-rupture and made hypertensive-hypervolaemic therapy safer. In delayed surgery, only the referring hospitals had to deal with this problem while the patients were waiting for surgery.

With the introduction of transcranial Doppler sonography (TCD) in 1982[1], a non-invasive, repeatable, and individual monitoring of the onset, the development, and the severity of vasospasm[1, 17, 18, 39, 40] became possible. This information allows better analysis of secondary deficits and a more specific haemodynamic treatment[14, 15].

The early removal of CSF contaminated with blood and blood clots, however, was not able to substantially reduce the rate of symptomatic vasospasm, which continued to occur in at least 10–20% of the patients[8, 19, 29, 36, 41, 42, 49]. By combining calcium antagonists[4, 6, 12, 15, 27, 35, 44] with the early surgery policy, the incidence of symptomatic vasospasm could be reduced to 3 to 13%.

Thus, the best management for ruptured aneurysm seems to be early aneurysm surgery in conjunction with the preventive administration of nimodipine and a Doppler-guided, selective, and preventive hypertensive therapy. With this regime, the total major morbidity ranges between 3 and 22% and the mortality between 2 and 11%[4, 5, 8, 12, 13, 27]. The incidence of severe disability or death due to vasospasm is between 1 and 7%[4–6, 12, 13, 27, 25].

There has been an increasing number of publications dealing with the beneficial effect of preventive nimodipine on DID after aneurysmal subarachnoid haemorrhage. These articles focus on the incidence and severity of ischaemic deterioration which was attributed to cerebral vasospasm or vasospasm plus complications. It seemed worthwhile to summarize the existing literature on clinical results with nimodipine and to determine whether there is sufficient evidence that the preventive use of nimodipine does improve the outcome.

Table 1. *Placebo Controlled Double-blind Prospective Randomized Studies*

Author	Nimodipine administration	Preoperative condition (H&H Grade)	Patients no.	Major morbidity	Mortality	DID	DID with severe disability or death
Allen et al.[3]	6 × 30 mg po 21 days	H&H I–II	56 N 60 P	36% 40%	5% 12%	23% 32%	2% 13%
Neil-Dwyer et al.[31]	6 × 60 mg po 21 days	H&H I–IV	25 N 25 P	20% 24%	4% 24%	? ?	16% 32%
Philippon et al.[34]	6 × 60 mg po 21 days	H&H I–III	31 N 39 P	58% 74%	? ?	7% 17%	10% 33%
Petruk et al.[33]	6 × 90 mg po 21 days	H&H III–V	73 N 83 P	14% 28%	48% 37%	36% 55%	? ?
Öhman + Heiskanen[32]	iv 2 mg/h 7–10 days + 6 × 60 po 11–14 days	H&H I–III	104 N 109 P	? ?	10% 14%	? ?	1% 8%

N: nimodipine group.
P: placebo group.

Placebo Controlled Double Blind Prospective Randomized Studies (Table 1)

Nimodipine was administered orally in the double-blind trials in which delayed surgical intervention was performed for ruptured aneurysms. The only exception was the study conducted by Öhman and Heiskanen[32], who administered nimodipine intravenously and operated early on approximately one third of their patients.

Allen et al.[3] in 1983 published a multi-centre prospective, double-blind, placebo-controlled randomized trial on the effect of oral nimodipine in aneurysmal subarachnoid haemorrhage patients. One hundred and twenty-five patients were submitted to the trial, nine were excluded because of protocol violations. Fifty-six received nimodipine at a dosage of 122–180 mg per day for 21 days. Treatment was started within 96 hours after aneurysm rupture. The patients were in grades I to II, and the operation had to take place within 14 days after the bleeding. At the end of the treatment period, 24 (43%) of the nimodipine and 28 (47%) of the placebo group demonstrated a good outcome. Three (5%) of the patients in the nimodipine group died, as did seven (12%) of the placebo group. One (1.8%) of the nimodipine group and eight (13.3%) of the placebo group developed delayed ischaemic deficits resulting in permanent disability or death. Serious side-effects were not observed, nor was there an increase in the rebleeding rate before surgery.

Since Allen et al. give no data concerning the overall clinical results and the surgical procedure, the overall result cannot be judged satisfactorily. The rate of secondary neurological deterioration is relatively high in the nimodipine group (23%) and even higher in the placebo group (32%).

In 1987, Neil-Dwyer[31] published a uni-centre double-blind placebo controlled prospective study on aneurysmal SAH patients. Seventy-five patients were proposed for the study and 25 were excluded because of protocol violations. Twenty-five of the remaining 50 received nimodipine and 25 placebo. Five patients were not operated on. Fifteen patients from each group were operated on within 21 days, a further five from each group after 21 days. The patients were in grades I to V. The treatment was started within 96 hours after the onset of SAH. The patients in the nimodipine group received 200 mg nimodipine intracisternally during the operation and 360 mg/day orally for 21 days after surgery. At follow-up three months later, 19 (76%) of the nimodipine group and 13 (52%) of the placebo group had a good outcome. Five (20%) and six (24%) patients, respectively, had remained in a poor condition, while one (4%) and six (24%), respectively, had died. The one death in the nimodipine group was caused by vasospasm and complications, as were three of the six fatal outcomes in the placebo group. Three additional deaths were due to vasospasm alone. In addition, three patients (12%) of the nimodipine group and two (8%)

of the placebo group experienced a poor outcome due to ischaemic deficits. Serious side-effects of nimodipine were not observed. The arterial blood pressure dropped about 5 mmHg and the CBF measured with Xe-133 decreased slightly over the 21 day treatment period.

Four patients suffered fatal rebleeding during treatment. Only one of these patients was from the nimodipine group, the other three belonged to the placebo group.

When analyzing the total group of 75 "intended to treat" patients, no change in the outcome of either group could be detected.

Phillipon[34] in 1986 presented a prospective randomized double blind study conducted in patients with H & H grade I–III following aneurysmal subarachnoid haemorrhage. Eighty-one patients were submitted for the trial, 11 were excluded due to protocol violations. Patients with intracerebral haematoma, hydrocephalus, and early surgery were not included. The surgical interventions took place after the fourth day post-aneurysm rupture. Nine patients in the nimodipine and 13 patients in the placebo group were not operated on. Thirty-one patients received nimodipine at a dosage of 360 mg/day for 21 days, while 39 received placebo. The treatment started within 72 hours. In both groups antifibrinolytics/transexamic acid were administered. The morbidity was 58% in the treatment group and 74% in the placebo group. Three patients (10%) of the nimodipine group suffered neurological dysfunctions with severe disability or death, as opposed to 13 (33%) of the placebo group. Five patients (16%) of the nimodipine group developed deficits due to vasospasm and other complications, as did 14 (36%) of the placebo group.

Phillipon's article does not contain an outcome analysis. The relatively high incidence of disability and death cannot be definitively analyzed on the basis of the available data. This was the only study in which the patients were additionally treated with antifibrinolytics.

In 1988, Taquoi et al.[43] published a meta-analysis of the material of the three double-blind oral trials[3, 31, 34], including the patients who had been excluded. In the nimodipine group the relative risk of mortality and severe morbidity due to vasospasm alone was significantly decreased. It was four times greater in the placebo group than in the treatment group (14.7 vs 2.8%). The risk of death or severe disability due to vasospasm and other complications was also significantly lower in the treatment group (13.4 versus 22.4%). Severe

morbidity and mortality, regardless of the cause, were also lower in the nimodipine group (26.8 and 35.7%).

Petruk[33] in 1988 presented a multi-centre randomized double-blind study on poor-grade (H & H III–V) aneurysmal subarachnoid haemorrhage patients in whom high doses of oral nimodipine were administered. One hundred and eighty-eight patients entered into the trial, 32 were excluded. Seventy-three of the valid cases received daily doses of 540 mg nimodipine orally for 21 days, 83 received placebo. Twenty-five of the nimodipine and 32 of the placebo group were in grade III, 34 and 49, respectively, in grade IV, and 14 and 12, respectively, in grade V. At follow-up 3 months later, 29% of the nimodipine group and 10% of the placebo group had made a good recovery. Ten percent and 24%, respectively, were in fair condition, while 14 and 27% were still in poor condition. Forty-seven percent of the nimodipine group and 39% of the placebo group died within 3 months. Eight patients (11%) of the nimodipine group and 25 (31%) of the placebo group suffered delayed ischaemic deficit due to vasospasm alone, 18 (25%) and 21 (26%), respectively, due to vasospasm and other complications.

Öhman and Heiskanen[32] in 1987 presented a placebo controlled randomized double-blind study on aneurysmal subarachnoid haemorrhage patients with intravenous administration of nimodipine. Two hundred and fifteen patients were enrolled in the study, two were excluded. One hundred and four received 48 mg/day intravenous nimodipine for 7 to 10 days starting after the diagnosis of ruptured aneurysm had been established, and then 360 mg orally per day for 21 days; 109 received placebo. The patients were in grades I to III and were operated on within 2 weeks, 58 within 72 hours, 69 between day 4 and 7, and 74 on day 8 or later following prospective stratification. Twenty-two patients died before surgery. Eight in the placebo group and four in the nimodipine group died as a result of rebleeding. This difference was statistically not significant. Two in the placebo group died of surgical complications and one in the treatment group for other reasons. One of the treated patients (1%) and nine (8%) from the placebo group died as a result of vasospasm. The analysis of the 108 (56N, 52P) patients operated on within 7 days showed that 90% of the treatment group and 80% of the placebo group had made a favourable outcome (good and fair). Eight percent of the treatment group and 8% of the placebo group were in poor condition, while 2% of the nimodipine and 12% of the placebo group died.

Table 2. *Uncontrolled Prospective Studies.* Uni-centre consecutive series with early surgery

Author	Nimodipine administration	Preoperative condition (H&H Grade)	Patients no.	Major morbidity	Mortality	DID	DID with severe disability or death
Auer[4]*	14 days i.v.	I–IV	65	14%	11%	6%	2%
	7 days p.o.	(I–III)	(47)	(4%)	(2%) ·	(6%)	(—)
Auer[6]* + Schneider	14 days i.v. 7 days p.o.	I–IV (I–III)	100 (74)	7% (7%)	17% (3%)	4% (4%)	1% (—)
Ljunggren** et al.[27]	14 days i.v.	I–IV (I–III)	60 (56)	22% (20%)	2% (2%)	3% (3%)	2% (2%)
Säveland** et al.[35]	14 days i.v.	I–IV (I–III)	100 (85)	22% (15%)	7% (7%)	7% (—%)	7% (6%)
Gilsbach*** et al.[12]	14 days i.v.	I–IV (I–III)	100 (72)	9% (3%)	9% (4%)	9% (13%)	1% (1%)

* Graz.
** Lund.
*** Freiburg.
( ) Subgroup of good condition patients (H&H I–III).

Uncontrolled Prospective Studies

Consecutive Uni-centre Series with Early Surgery and Intravenous Nimodipine (Table 2)

Öhman and Heiskanen's study is the only double-blind trial on early surgery and intravenous nimodipine. Apart from this only prospective, non-controlled uni-centre or multi-centre series, and a double-blind multi-centre dose comparison study have been published.

Auer[4] was the first to publish a uni-centre open prospective study in 1984 on patients who were consecutively operated on early and treated preventively with nimodipine. Sixty-five patients were operated on within 72 hours after aneurysm rupture. They were preoperatively in clinical grades I to V (47 patients in grades I to III and 18 in grades IV to V). The cisterns were intraoperatively rinsed with a nimodipine solution $(2.4 \times 10^{-5} M)$. The intravenous nimodipine treatment was initiated during the operation. The patients received 24 to 48 mg/h nimodipine for 14 days and 240 mg daily for the next 7 days. At the six months follow-up, 75% of the patients were in good condition, 6% in fair, and 7% in poor condition. Eleven percent had died. In the good condition subgroup 94% had made a good recovery, 4% fair recovery, and 2% had died. Two patients (3%), who preoperatively had been in grade I and III, respectively, developed transient delayed ischaemic deficits, while two additional (3%)

patients – one in grade I and one in grade IV – suffered permanent neurological deficit. In only one of them (2%), however, did severe disability ensue. In none of the patients were secondary dysfunctions the result of vasospasm alone; a combination of vasospasm and other complications was responsible for all of the dysfunctions.

The study published by Auer and Schneider[6] in 1986 was based on further experience with a total of 100 patients treated with the same regime. The results were principally the same as in the Auer article[4]: in the entire group of grades I to V patients, 7% experienced major morbidity, 17% died, and 1% suffered delayed dysfunction with severe disability. In the good condition subgroup, 7% of the patients were disabled, 4% died, and none had a DID with subsequent severe disability.

In 1984, Ljunggren et al.[27] presented a comparable uni-centre prospective series on 60 consecutive patients operated on within five days after aneurysm rupture. They were preoperatively in grades I to V (56 in grades I to III and 4 in grades IV and V). As in the Auer study, nimodipine was topically applied during surgery and the intravenous treatment was started intraoperatively at a dosage of 48 mg/d for 14 days. At follow-up from 2 months to 1.5 years, 77% of the patients were in good condition, while 15% remained in fair and 7% in poor condition. Two percent of the patients had died. In the group of good condition patients (grades I to III) 79% had made a good recovery,

Table 3. *Uncontrolled Open Prospective Studies.* Multi-centre consecutive series with early surgery

Author	Nimodipine administration	Preoperative condition (H & H Grade)	Patients no.	Major morbidity	Mortality	DID	DID with severe disability or death
Auer* et al.[5]	topical + 14 days i.v. 48 mg/d 7 days 360 mg/d p.o.	I–III	120	3%	3%	7%	2%
Gilsbach** et al.[13]	topical + 9–14 days 48/72 mg/day	I–V (I–III)	204 (162)	11% (6%)	11% (7%)	4% (4%)	2% (2%)

* Berlin, Bern, Freiburg, Graz.
** Berlin, Bern, Bielefeld, Freiburg, Graz, Linkoeping, Lund, Stockholm, Umea, Wien.
( ) Subgroup of good condition patients (H & H I–III).

13% a fair, and 7% a poor recovery; 2% had died. Two (3%) of the 60 patients, both preoperatively in grade II, suffered delayed ischaemic deficits. One recovered to a fair condition and one died after a large cerebral infarction (caused by pneumonia!). This corresponds to a 2% rate of DID with severe disability or death. This low rate of deficits is in contrast with the incidence and severity of delayed angiographic vasospasm, which was not substantially reduced during nimodipine treatment.

Säveland et al.[35] in 1986 reported the results of 100 patients who were treated in the same centre and according to the same protocol. Eighty-five patients were in grades I to III and 15 in grades IV to V. At follow-up 3 months to 3 years after surgery, 71% had made a good, 11% a fair, and 11% a poor recovery, while 7% of the patients had died. In the good condition subgroup 79% made a good, 8% a fair, and 7% a poor recovery; 6% had died. Two percent of the patients had suffered a temporary delayed ischaemic deficit, while 5% experienced secondary deterioration with severe disability in 2% and death in 3%.

Gilsbach et al.[12] using the same protocol published the results of one-hundred consecutive patients in a non-controlled uni-centre prospective study in 1987. Seventy-two were in good preoperative condition and 28 in poor condition (H & H grade IV–V). The aneurysms were surgically treated within 72 hours after the rupture. The treatment was started intraoperatively with a topical application of nimodipine on the exposed arteries and intravenous administration of 48 mg/d, which was maintained until the 14th day after surgery. Thereafter, some patients received nimodipine at an oral dosage of 240 mg/day until day 21. At the six

month follow-up, 72% of the patients were in a good, 10% in a fair, and 9% in poor condition. Nine percent of the patients had died. In the good condition subgroup 83% had made a good recovery, 10% a fair, and 3% a poor recovery; 4% had died. Twelve patients suffered delayed neurological deterioration: four from vasospasm alone and five from vasospasm and other complications. In three patients the secondary deficit was not related to vasospasm. Only one (1%) of the patients preoperatively in grade III died as a result of vasospasm and other complications, none suffered permanent disability.

Consecutive Multi-centre Series with Early Surgery and Intravenous Nimodipine (Table 3)

Auer et al.[5] in 1986 published the first multi-centre open prospective trial on nimodipine and early aneurysm surgery. One hundred and twenty patients in good preoperative condition (H & H I–III) were evaluated. In order to clearly recognize the development of secondary neurological deterioration, grade IV and V patients as well as patients with preoperative neurological deficit were excluded. The patients were operated on within 72 hours after the last bleed (in grade III patients within 48 hours). Intraoperatively, nimodipine was applied topically and the intravenous administration of nimodipine (48 mg/day) was started and maintained until day 14. Then the patients received 240 to 270 mg orally until day 21. At the six month follow-up, 90% were in good, 4% in fair, and 3% in poor condition; 3% had died. Eight (7%) patients experienced transient and two (2%) permanent delayed ischaemic deteriorations. One of the two, both preopera-

Table 4. *Uncontrolled Prospective Studies.* Uni-centre series with early and delayed surgery

Author	Nimodipine administration	Preoperative condition (H&H Grade)	Patients no.	Morbidity + mortality	DID		
					transient	permanent	lethal
Grotenhuis and Bettag[16]	14 days 48 mg/d i.v. 4 days 240/d p.o.	H&H I–IV (I–III)	61	26% (13%)	3% (3%)		2%

( ) Subgroup of good condition patients (H&H I–III).

Table 5. *Uncontrolled Open Prospective Studies.* Multi-centre series with early and delayed surgery or without surgery

Author	Nimodipine administration	Preoperative condition (H&H Grade)	Patients no.	Major morbidity	Mortality	DID with severe disability or death
Kazner et al.[22]	i.v. ≥ 7 days	I–III	104	14%	6%	11%
Kazner[23]	7–14 days i.v. 4–6 days p.o.	I–III	284	10%	7%	7%

tively in grade I, remained in poor and one in fair condition, making a 1% rate of delayed neurological dysfunctions with severe disability.

In 1987, Gilsbach et al.[13] presented results of a similar multi-centre group. They conducted a double-blind dose comparison multi-centre study which used the same protocol, testing a dosage of 48 mg/day against 72 mg/day. Two hundred and thirty-seven aneurysmal subarachnoid haemorrhage patients were submitted. Thirty-three were excluded due to protocol violations. One hundred and sixty-two were in grades I to III and 42 in grades IV to V. At the six month follow-up, 32% of the grade IV and V patients had made a good recovery, 5% a fair, 11% a poor recovery; 11% had died. In the subgroup of good condition patients 83% had made a good, 5% a fair, 6% a poor recovery, while 7% had died. Seven (3%) patients suffered ischaemic deficits which in three (1%) patients led to severe disability and in one (0.5%) to death. This corresponds to a 3% rate of severe disability or death following delayed neurological dysfunction. In only one of these three cases could the severe disability be attributed to vasospasm alone; in the two others it was caused by vasospasm and intraoperative complications. There was no significant difference with regard to the outcome and side effects between the two dosage groups. The morbidity and mortality, however, were slightly lower in the high dosage group.

Uni-centre Series with Early and Delayed Surgery (Table 4)

Grotenhuis and Bettag[16] in 1986 published an article dealing with 61 consecutive patients in grades I to IV admitted within 6 days after the aneurysm rupture. The treatment was started after the diagnosis of subarachnoid haemorrhage had been made. Forty-eight mg nimodipine was administered intravenously per day for at least 14 days. Thereafter, 240 mg/day were given orally for a minimum of 4 days. Seventy-four percent of the patients made a favourable outcome and 26% died or had a poor outcome. In the good condition subgroup of 47 patients, 87% had a favourable and 13% an unfavourable outcome. Two patients (3%) who preoperatively had been in grade I to III suffered temporary delayed ischaemic deficits. One patient (2%) died; he had been operated on the day of the rupture and developed an infarction 30 hours later.

Serious adverse reactions were not observed. In three patients the blood pressure dropped. In two of them it returned to normal after the dosage had been reduced.

Open Multi-centre Series with Early and Delayed Surgery or Without Surgery (Table 5)

Kazner[23] published an open prospective multi-centre study in 1986 on 284 subarachnoid haemorrhage patients, 240 of whom underwent aneurysm operation. In 44 patients there was either no aneurysm or operation was not possible. The patients were in grades I to III and the treatment consisted of intravenous administration of nimodipine (48 mg/day) for 7 to 14 days followed by oral administration of 240 to 360 mg/day for 4 to 10 days. Eighty-four percent had a favourable outcome (good and fair), while 10% remained in a poor condition and 7% died. Ten patients (4%) suffered delayed ischaemic deficits with severe disability or death due to vasospasm. In addition, nine patients (3%) developed deficits due to vasospasm and other complications. No serious side-effects were observed. In only 3% of the patients did the treatment have to be interrupted or terminated because of side-effects.

A previous publication by the same author and group[22] on 104 patients showed generally the same results.

Discussion

Uncontrolled, open trials in general reflect the problem of the evaluation and comparison of a new pharmacological treatment. In the uncontrolled studies the favourable results obtained with preventive nimodipine administration could partially be attributed to a more careful supervision, more surgical experience, and improved management qualities. In other words, the improvement of the results may be due to the drug effect and/or due to such, as yet unidentified factors. Therefore, the comparison of recent studies with results from previous time periods, even if well-documented, is always problematical.

The remarkable reduction of DID in patients treated intravenously with nimodipine and early aneurysm surgery[4, 12, 27], in comparison with previous results without nimodipine, was one of the reasons why some authors decided not to conduct placebo-controlled studies after the first consecutive uncontrolled trials had been published. The better the results became, the more difficult it was to justify the necessity of a double-blind study (at least with intravenous administration of nimodipine in early aneurysm surgery patients). That means that the question as to what extent the patients benefit from preventive nimodipine alone cannot be answered definitively for early surgery and intravenous drug application. The double-blind study by Öhman and Heiskanen on intravenous treatment and early surgery confirmed the low incidence of DID, but it was not detailed and extensive enough to convincingly demonstrate the superiority of early surgery with nimodipine over early surgery without nimodipine.

In contrast, the trials of the effect of oral nimodipine and delayed surgery were primarily performed in a double-blind fashion. The effect was obviously positive, but only in small series with a delayed treatment regime, which produces per se more unfavourable outcomes and ischaemic deficits than early surgery.

The aim of the above-mentioned trials was to evaluate the effect of nimodipine on the development of delayed ischaemic deficits. These secondary deteriorations were originally thought to be mainly a consequence of hypoperfusion due to the reduced vessel diameter. The detailed analysis of the results, however, shows that in many patients a combination of vasospasm and other, mostly surgical complications was responsible for the development of new deficits[4, 13, 14, 22, 28, 33–35]. In both instances the significance of symptomatic vasospasm for secondary (ischaemic) neurological deficits cannot be clearly defined because the diagnosis vasospasm was mainly based on the clinical course and rarely on (routine) control angiography or TCD. Therefore, suspected causes of secondary (ischaemic) deterioration in earlier publications must be judged carefully. With the availability of the transcranial Doppler sonography, it is now possible to monitor non-invasively and continuously the development of cerebral vasospasm with a high degree of accuracy[1, 2, 17, 18, 40] and to discriminate better spasm-related deficits from operative deficits. Thus, a more exact assessment of vasospasm in the development of delayed neurological dysfunction should be possible in the future. When considering the effect of a new drug on the sequelae of SAH, the reduction of delayed ischaemic deficits alone should not be overemphasized, unless the overall results can be positively influenced. An improvement in the treatment regime, including the prevention of ischaemic deficit, should be measurable in the overall outcome with a decreased morbidity and mortality. Indeed, the patients preventively treated with nimodipine present results superior to those related without nimodipine. It is, however, not possible to decide whether the improved overall outcome associated with nimodipine was caused by an anti-vasospastic effect, a cytoprotective effect, and/or other unidentified factors such as improved general supervision and management.

Controlled Studies

The results of the controlled and double-blind studies, which with one exception apply the oral form of nimodipine administration, show that severe disability or death due to ischaemic deterioration is lower in the nimodipine than in the placebo group (2 to 16% versus 13 to 33%)[3, 31-34]. The same holds good for major morbidity, which is lower in the nimodipine group (14 to 58% versus 24 to 74%). These findings strongly indicate that nimodipine reduces the occurrence of severe neurologic deficit and death from cerebral vasospasm. The studies with delayed surgery also reflect the disadvantage of the delayed policy: a relatively high rate of rebleeding, a mortality of 4 or 10% in grade I–III (H & H) and a major morbidity of 36 to 58% which are higher than after early surgical intervention. Even the incidence of DID with fixed neurological deficit in the nimodipine group is higher than after early surgery (see Tables 1 and 2).

The trials of orally treated patients are small with a high rate of exclusions and the available data are not detailed enough to analyze the problems associated with surgery, bleeding, and rebleeding. The exclusion per se had no influence on the results, as an analysis of all patients (including the excluded ones) by Taquoi et al.[43] confirmed. This meta-analysis of three studies[3, 31, 34] confirmed that nimodipine given as preventive oral therapy is effective in preventing death or severe neurological deficit.

Open Uni- and Multi-centre Studies

The uncontrolled and open uni- and multi-centre studies using topical and intravenous nimodipine in combination with early aneurysm surgery[4, 6, 12, 27, 35] show that this treatment regime yields favourable results. The outcome in the different studies was very similar, as was the incidence of delayed ischaemic deficits. In the comparable subgroup of good condition patients (H & H I–III) between 78 and 94% of the patients made a good outcome, while 4 to 7% died. The incidence of delayed ischaemic deterioration with severe disability or death ranged between 1 and 7%. The similarity of the results of the various groups is striking. Apparently, local differences and local specializations do not play a major role. The conclusion seems to be justified that the results obtained with the combination of early aneurysm surgery and preventive intravenous nimodipine (including topical nimodipine administration?) are superior to those obtained after oral nimodipine treatment and delayed operation. The comparison with previous trials (historical controls) which is, however, of limited value shows that the incidence of delayed ischaemic deficits has been reduced substan-

tially. In several series of early operation but without calcium channel blockers the incidence of permanent dysfunction after delayed ischaemic deterioration was between 13 and 20%[8, 21, 25, 26, 42, 45, 49]. Without a well-documented prospective double-blind study it is, however, not possible to define the extent to which nimodipine or other factors (surgery, postoperative management) are responsible for the good results. It is worth noting that comparable results without nimodipine prevention were recently reported, where vasospasm was treated after the onset of symptoms by induced hypervolaemia and hypertension alone[7, 10]. This type of treatment, however, necessitates intensive monitoring and is associated with a relatively high rate of cardiopulmonary complications[7, 10]. Even if the results of the two treatment regimes are comparable, early surgery with preventive nimodipine offers the simpler management regimen.

References

1. Aaslid R, Markwalder T-M, Nornes H (1982) Noninvasive transcranial Doppler ultrasound of flow velocities in basal cerebral arteries. J Neurosurg 57: 769–774
2. Aaslid R, Huber P, Nornes H (1984) Evaluation of cerebrovascular spasm with transcranial Doppler ultrasound. J Neurosurg 60: 37–41
3. Allen GS, Hyo SA, Preziosi TJ, Battye R, Boone SC, Chou SN, Kelly DL, Weir BK, Crabbe RA, Lavik PJ (1983) Cerebral arterial spasm – a controlled trial of nimodipine in patients with subarachnoid haemorrhage. N Engl J Med 308: 619–624
4. Auer LM (1984) Acute operation and preventive nimodipine improve outcome in patients with ruptured cerebral aneurysms. Neurosurgery 15: 57–66
5. Auer LM, Brandt L, Ebeling U, Gilsbach J, Groeger U, Harders A, Ljunggren B, Oppel F, Reulen HJ, Säveland H (1986) Nimodipine and early aneurysm operation for good condition SAH patients. Acta Neurochir (Wien) 82: 7–13
6. Auer LM, Schneider GH, Auer T (1986) Computerized tomography and prognosis in early aneurysm surgery. J Neurosurg 65: 217–221
7. Awad IA, Carter P, Spetzler RF, Medina M, Williams FW (1987) Clinical vasospasm after subarachnoid haemorrhage: Response to hypervolaemic haemodilution and arterial hypertension. Stroke 18: 365–372
8. Brandt L, Ljunggren B, Säveland H (1986) Prophylaxe ischämischer neurologischer Defizite nach Subarachnoidalblutung. In: Kazner E (ed) Nimotop in der Prophylaxe und Therapie neurologischer Ausfälle durch zerebrale Ischämie. Krankenhausarzt. G Braun, Karlsruhe, pp 67 – 76
9. Ecker A, Riemenschneider PA (1951) Arteriographic demonstration of spasm of the intracranial arteries with special reference to saccular arterial aneurysms. J Neurosurg 8: 660–667
10. Finn SS, Stephensen SA, Miller CA, Drobnich L, Hunt WE (1986) Observations on the perioperative management of aneurysmal subarachnoid haemorrhage. J Neurosurg 65: 48–62
11. Fisher CM, Roberson GH, Ojemann RG (1977) Cerebral va-

sospasm with ruptured saccular aneurysm – the clinical manifestations. Neurosurgery 1: 245–248

12. Gilsbach J, Harders A, Hornyak ME (1987) Cerebral vascular spasm in aneurysm surgery and its clinical significance. In: Sinha KK, Chandra P (eds) Progress in clinical neurosciences. Catholic Press, Ranchi, pp 125–133

13. Gilsbach JM, Ljunggren B, v Holst H, Seiler R, Mokry M, v Essen C, Conzen M (1987) Outcome in 204 SAH patients subjected to early aneurysm surgery and intravenous nimodipine. Report of a multicenter double blind dose comparison study. In: Prevention and treatment of delayed ischaemic dysfunction in patients with subarachnoid haemorrhage. An update. Barcelona, Sept. 7

14. Gilsbach JM, Harders A, Hornyak ME (1988) Does vasospasm cause major morbidity and mortality in early aneurysm surgery? In: Proceedings on cerebral vasospasm, Charlottesville, 1987, Raven Press (in press)

15. Gilsbach JM, Harders AG, Eggert HR, Hornyak ME (1988) Early aneurysm surgery: a 7 years clinical practice report. Acta Neurochir (Wien) 90: 91–102

16. Grotenhuis JA, Bettag W (1986) Prevention of symptomatic vasospasm after SAH by constant venous infusion of nimodipine. Neurol Res 8: 243–248

17. Harders A (1986) Neurosurgical applications of transcranial Doppler sonography. Springer, Wien New York

18. Harders AG, Gilsbach JM (1987) Time course of blood velocity changes related to vasospasm in the circle of Willis measured by transcranial Doppler ultrasound. J Neurosurg 66: 718–728

19. Kassell NF, Boarini DJ, Adams HP, Sahs AL, Graf CJ, Torner HC, Gerk MK (1981) Overall management of ruptured aneurysm: comparison of early and late operation. Neurosurgery 9: 120–128

20. Kassell NF, Peerless SJ, Durward QJ, Beck DW, Drake CG, Adams HP (1982) Treatment of ischaemic deficits from vasospasm with intravascular volume expansion and induced arterial hypertension. Neurosurgery 11: 337–343

21. Kassell NF, Torner JC (1984) The international cooperative study on timing of aneurysm surgery – an update. Stroke 15: 556–570

22. Kazner E, Sprung Ch, Adelt D, Ammerer HP, Karnick P, Baumann H, Böker DK, Grotenhuis JA, Jaksche H, Istaitih A-R (1985) Clinical experience with nimodipine in the prophylaxis of neurological deficits after subarachnoid haemorrhage. Neurochirurgia 28: 110–113

23. Kazner E (1986) Ergebnisse einer multizentrischen Studie mit Nimodipin in der Prophylaxe ischämischer neurologischer Defizite nach Subarachnoidalblutungen. In: Kazner E (ed) Nimotop in der Prophylaxe und Therapie neurologischer Ausfälle durch zerebrale Ischämie. Krankenhaus Arzt. G Braun, Karlsruhe, pp 67 – 76

24. Kågström E, Greitz T, Hanson J, Galera R (1966) Changes in cerebral blood flow after subarachnoid haemorrhage. In: Proceedings of the Third International Congress of Neurological Surgery, 1965. Int Congr Series No. 110. Excerpta Medica, Amsterdam, pp 629–633

25. Ljunggren B, Brandt L, Sundbärg G, Säveland H, Cronquist S, Stridbeck H (1982) Early management of aneurysmal subarachnoid haemorrhage. Neurosurgery 11: 412–418

26. Ljunggren B, Säveland H, Brandt L (1983) Causes of unfavourable outcome after early aneurysm operation. Neurosurgery 13: 629–633

27. Ljunggren B, Brandt L, Säveland H, Nilsson P-E, Cronquist S, Andersson KE, Vinge H (1984) Outcome in 60 consecutive patients treated with early aneurysm operation and intravenous nimodipine. J Neurosurg 62: 864–873

28. Mee EW, Dorrance DE, Low D, Neil-Dwyer G (1986) Cerebral blood flow and neurological outcome: a controlled study of nimodipine in patients with subarachnoid haemorrhage. J Neurol Neurosurg Psychiatry 49: 469

29. Milhorat TH, Krautheim M (1986) Results of early and delayed operation for ruptured intracranial aneurysms in two series of 100 consecutive patients. Surg Neurol 26: 123–128

30. Muizelaar JP, Becker DP (1986) Induced hypertension for the treatment of cerebral ischaemia after subarachnoid haemorrhage. Direct effect on cerebral blood flow. Surg Neurol 25: 317–325

31. Neil-Dwyer G, Mee E, Dorrance D, Lowe D (1987) Early intervention with nimodipine in subarachnoid haemorrhage. European Heart J 8: 41–47

32. Öhman J, Heiskanen O (1987) The effect of nimodipine on cerebrovascular spasm after subarachnoid haemorrhage – a randomized, double-blind study, presented at the 2nd World Congress of Neuroscience, Budapest, Hungary, August 16–21, 1987

33. Petruk KC, West M, Mohr G, Weir BKA, Benoit BG, Gentili F, Disney LB, Khan MI, Grace M, Holness RO (1988) Nimodipine treatment in poor grade aneurysm patients. Result of a multicentre, double-blind, placebo controlled trial. J Neurosurg 68: 505–517

34. Philippon J, Grob R, Dagreou F, Guggiari M, Rivierez M, Viars P (1986) Prevention of vasospasm in subarachnoid haemorrhage. A controlled study with nimodipine. Acta Neurochir (Wien) 82: 110–114

35. Säveland H, Ljunggren B, Brandt L, Messeter K (1986) Delayed ischaemic deterioration in patients with early aneurysm operation and intravenous nimodipine. Neurosurgery 18: 146–150

36. Saito I, Yasuichi U, Sano K (1977) Significance of vasospasm in the treatment of ruptured intracranial aneurysms. J Neurosurg 47: 412–429

37. Saito I, Sano K (1979) Vasospasm following rupture of cerebral aneurysm. Neurol Med Chir (Tokyo) 19: 103–107

38. Sano K, Saito I (1978) Timing and indication of surgery for ruptured intracranial aneurysms with regard to cerebral vasospasm. Acta Neurochir (Wien) 41: 49–60

39. Seiler RW, Grolimund P, Aaslid R, Huber P, Nornes H (1986) Cerebral vasospasm evaluated by transcranial ultrasound with clinical grade and CT-visualized subarachnoid haemorrhage. J Neurosurg 64: 594–600

40. Seiler RW, Grolimund P, Zurbruegg H (1987) Evaluation of the calcium-antagonist nimodipine for the prevention of vasospasm after aneurysmal subarachnoid haemorrhage. A prospective transcranial Doppler ultrasound study. Acta Neurochir (Wien) 83: 7–16

41. Sundt TM, Shigeaki K, Fode NC, Whisnant JP (1982) Results and complications of surgical management of 809 intracranial aneurysms in 722 cases: Related and unrelated to grade of patient, type of aneurysm, and timing of surgery. J Neurosurg 56: 753–765

42. Taneda M (1982) Effect of early operation for ruptured aneurysms on prevention of delayed ischaemic symptoms. J Neurosurg 57: 622–628

43. Taquoi G, Porto L, Tettenbron D (1988) Nimodipine oral treatment in subarachnoid haemorrhage – Overview of results in

mortality and severe morbidity from three randomized controlled trials. Ann NY Acad Sci

44. Tettenborn D, Porto L, Ryman T, Strugo V, Taquoi G, Battye R (1987) Survey of clinical experience with nimodipine in patients with subarachnoid haemorrhage. Neurosurg Rev 10: 77–84

45. Vapalahti M, Ljunggren B, Säveland H, Herniesniemi J, Brandt L, Tapaninaho A (1984) Early aneurysm operation and outcome in two remote Scandinavian populations. J Neurosurg 60: 1160–1162

46. Wilkins RH (1980) Attempt at prevention and treatment of intracranial arterial spasm: a survey. In: Wilkins RH (ed) Cerebral arterial spasm. Williams & Wilkins, Baltimore, pp 542–555

47. Wilkins RH (1980) Cerebral arterial spasm. Williams & Wilkins, Baltimore

48. Wilkins RH (1986) Attempts at prevention and treatment of intracranial arterial spasm: an update. Neurosurgery 18: 146–150

49. Yamamoto I, Hara M, Ogura K, Suzuki Y, Nakane T, Kageyama N (1983) Early operation for ruptured intracranial aneurysms: Comparative study with computed tomography. Neurosurgery 12: 169–174

Correspondence: Prof. Joachim M. Gilsbach, Neurochirurgische Universitätsklinik, Hugstetter Strasse 55, D-7800 Freiburg i. Br., Federal Republic of Germany.

Acta Neurochirurgica, Suppl. 45, 51–55 (1988)
© by Springer-Verlag 1988

Review of Treatment of Symptomatic Cerebral Vasospasm with Nimodipine

F. Buchheit and **P. Boyer**

Neurosurgical Unit, Hautepierre Hospital, Strasbourg, France

Summary

A number of randomized studies have shown the efficacy of nimodipine, administered either orally or intravenously, for the prevention of vasospasm and its clinical consequences in patients with subarachnoid haemorrhage following rupture of an intracranial aneurysm.

It remained to be proven that nimodipine could also act on already established vasospasm. This was the aim of a multicentre study carried out in France between 1984 and 1986.

Of a total of 127 patients with known clinically and/or angiographically diagnosed vasospasm, 73 (group N) underwent intravenous treatment with nimodipine and 54 (group P) with placebo within 24 hours of the onset of vasospasm. There was shown to be a significant reduction in mortality and morbidity in group N (33%) compared with group P (52%). Where vasospasm was the sole determining factor (63% of all patients), the decrease in mortality and severe morbidity rate was even greater in group N (11%) compared with group P (31.5%).

These results can be viewed as confirmation of the efficacy of nimodipine in treating the ischaemic consequences of established vasospasm.

Keywords: Cerebral vasospasm; subarachnoid haemorrhage; treatment; nimodipine.

Introduction

It has become increasingly clear in recent years that the significant advances made in the field of microsurgery of ruptured intracranial aneurysms has not had a radical effect on the overall prognosis for this condition. Mortality and neurological deficits are still frequent despite the high standard of surgical technique. New therapeutic approaches have therefore been adopted towards the two principal factors responsible for the poor prognosis: rebleed and vasospasm.

As a result of in-depth studies on rebleeding, its incidence is now estimated at 12 to 30% during the first two weeks after rupture of the aneurysm[10, 11, 12, 19]. It has also emerged that rebleeding during the first few days is more common than previously thought. In order to avert this "early" risk factor, several neurosurgeons have therefore suggested early surgery for ruptured aneurysms within 72 hours of the subarachnoid haemorrhage[3, 15, 17, 21, 22, 23].

Similarly, knowledge of vasospasm and its likely time of onset has also improved. Secondary vasospasm occurs in 20 to 50% of patients, establishing itself from day 3 onwards and occurring most commonly on day 8 . This pattern of vasospasm onset has become an additional argument in favour of early surgery with the aim both of preventing rebleeding and, to a certain extent, of reducing the risk of vasospasm by means of cisternal lavage during surgery in the hope of eliminating some of the spasmogenic substances resulting from the subarachnoid haemorrhage. In the latter respect the hopes have proved unfounded, since many patients still develop vasospasm and ischaemic neurological complications[15, 23].

Meanwhile, following the failure of all conventional agents thought likely to act on the muscle in vessel walls, pharmacological research has concentrated on studying calcium antagonists and their effect on cerebral vasospasm. Of the different preparations investigated, nimodipine has been shown to exhibit cerebral selectivity.

It has now been suggested that nimodipine could be used to prevent vasospasm. A number of clinical trials, open and controlled, have shown its efficacy in preventing the neurological complications of vasospasm when oral or intravenous treatment is initiated within 96 hours of rupture of the aneurysm. The drug has also been administered intracisternally to patients undergoing early surgery.

Prophylactic Efficacy of Nimodipine

The prophylactic potential of nimodipine was first demonstrated by Allen *et al.*[1] in 1983 in a double-blind, randomized study. The aim of this study was to assess the number of deaths and severe neurological deficits from vasospasm after commencing prophylactic treatment with nimodipine (120–180 mg daily in 6 oral doses) within 96 hours of subarachnoid haemorrhage. In the active group (56 patients) there was only one poor outcome compared with 8 in the placebo group (60 patients).

A similar randomized study was performed by Philippon et al.[18]. Oral prophylactic treatment was commenced within 72 hours of aneurysmal rupture in patients in Hunt and Hess grades I to III. The dosage was higher than that used by Allen et al. (60 mg every 4 hours, i.e. 360 mg daily, over 21 days). 31 patients were analyzed in the active group (N) and 39 in the placebo group (P). Regular control angiograms showed a reduction in the incidence of arterial vasospasm in group N (52% of patients) compared with group P (72% of patients), although this finding is not statistically significant. On the other hand, clinical assessment on completion of treatment revealed a significant difference between the two groups in terms of the number of poor outcomes (deaths and severe deficits) due to vasospasm alone: 6.4% in group N compared with 25.6% in group P.

A British prophylactic study conducted by Mee et al.[16] involved 25 patients treated orally with 300 mg nimodipine daily and 25 patients treated with placebo within 96 hours of the haemorrhage. The study also incorporated intracisternal administration of nimodipine or placebo solution during surgery. 3 deaths due to vasospasm were reported in group P and none in group N.

It can be concluded from these three randomized studies that prophylactic treatment, commenced early before the onset of angiographically detectable vasospasm and its clinical consequences, significantly reduces the number of poor outcomes due to vasospasm, though without eliminating them completely even at high dosages. Battye[7], on the other hand, attempted to highlight the role played by dosage in prevention by randomizing three groups of patients (n = 55, 56 and 59) receiving oral doses of 30, 60 and 90 mg nimodipine respectively every 4 hours. The number of poor outcomes was higher in the group taking 30 mg (12.7%) than in the two other groups (5%).

A number of open studies have confirmed the unquestionable prophylactic efficacy of nimodipine using a different approach. These studies involved administering nimodipine intravenously (1–2 mg/hour) in conjunction with intracisternal administration during early surgery (within 72 hours of rupture of the aneurysm).

Such studies have been performed by Auer et al.[3, 4, 5, 6], Ljunggren et al.[14] and Saveland et al.[20]. Mortality ranged between 1.5 and 7% and the incidence of resulting severe neurological deficits between 2.5 and 5%. These figures are appreciably lower than those found in the literature in patients not treated with nimodipine. In particular, Ljunggren compares the 3% severe morbidity in his study with the 13% rate amongst 137 patients who underwent early surgery without nimodipine treatment.

Therapeutic Efficacy of Nimodipine

Management of ruptured intracranial arterial aneurysms does not always provide scope for prophylactic treatment of vasospasm or early surgery. In many cases patients present at neurosurgical units after day 4, having already suffered arterial spasm and possibly with established ischaemic symptoms. It was important to determine whether commencing nimodipine treatment at this stage could improve the prognosis and to demonstrate this in a randomized study. An open study performed by Koos et al.[13] suggested that there is a reduction in mortality and severe morbidity after i.v. treatment. This acted as an incentive to undertake the multicentre study carried out in France between 1984 and 1986 which is analyzed in detail in this paper.

1. Patients and Methods

A. Inclusion Criteria

The study was carried out at 13 French neurosurgical units*. Each centre randomized its own patients, all with subarachnoid haemorrhage after rupture of an intracranial arterial aneurysm and fulfilling either of the following criteria:

a) pre- or postoperative neurological deterioration in accordance with the Hunt and Hess scale where there was a sure or highly probable correlation with vasospasm after eliminating all other possible causes (rebleeding, hydrocephalus) by performing a CT scan, together with evidence of vasospasm in a number of patients on a control angiogram

b) evidence of severe and diffuse vasospasm on the preoperative cerebral angiogram

In each of these two categories intravenous treatment was commenced within 24 hours of the onset of clinical deterioration or of the angiographically confirmed spasm. A review board subsequently excluded all cases where these criteria were not met. In particular, any patients who developed postoperative neurological complications within 12 hours of surgery were excluded.

---

* Brest, Caen, Chamalières, Limoges, Lille, Marseille, Montpellier, Paris Salpêtrière, Poitiers, Rennes, Strasbourg, Toulouse Purpan, Tours.

B. Method of Treatment

The 188 patients recruited in this way underwent continuous i.v. treatment for at least 7 days. 102 patients (N) received an infusion of an alcohol-based 0.02% solution of nimodipine at a dose of 0.030 mg/kg/h (about 2 mg/hour for a person weighing 70 kg), while 86 patients (P) received only the alcohol-based solution at a rate of 10 ml/hour.

C. Evaluation

The patients' condition was evaluated at the end of i.v. treatment (12.2 days in group N and 11.4 days in group P) and again, in the case of survivors, on leaving the neurosurgical unit (34.9 days in group N and 36.9 days in group P). They were classified in accordance with the Glasgow Outcome Scale: good outcome (grades I and II), severe disability (groups III and IV) or death. The aim of the study was to assess the number of deaths and severe deficits attributable to vasospasm alone.

2. Results

After exclusion by the review board of 61 patients who did not fully satisfy the entry criteria, 127 cases remained eligible for evaluation (73 in group N and 54 in group P).

There was no significant difference between the two groups in terms of demographic data (Table 1), site of aneurysms (Table 2) and clinical state in accordance with the Hunt and Hess grades of entry into the study (Table 3). The extent of the haemorrhage on the CT scan at the time of entry was comparable in the two groups. The mean interval between haemorrhage and

Table 1. *Demographic Data*

		Nimodipine group N = 73	Placebo group N = 54
Age		48.5 ± 14.3	47.3 ± 15.4
Sex	M	33	28
	F	40	26
Weight		68.0 ± 14.8	70.6 ± 13.2
Height		167.5 ± 8.7	168.0 ± 8.0

Table 2. *Location of Ruptured Aneurysm*

	Nimodipine N = 73	Placebo N = 54
Internal carotid artery	16 (21.9%)	9 (16.7%)
Middle cerebral artery	20 (27.4%)	10 (18.6%)
Anterior communicating artery	32 (43.8%)	26 (48.1%)
Others	5 (6.8%)	9 (16.7%)

Table 3. *Hunt and Hess Grade on Entry Into Study*

	Nimodipine N = 73	Placebo N = 54
I–II	12 (16.4%)	12 (22.2%)
III	26 (35.6%)	21 (38.9%)
IV	35 (47.9%)	21 (38.9%)

surgery was 13 ± 11 days in group N and 12 ± 12 days in group P.

On final evaluation of the case reports, a sub-group was formed for patients whose outcome was attributable to vasospasm alone. This sub-group comprised 80 (63%) of the 127 patients (N = 42, P = 35). In the case of the remaining patients, the outcome was attributable to vasospasm and some other complication (rebleeding, surgical complications, etc.).

This made it possible to assess the overall outcome for the entire group of 127 patients and to evaluate separately the sub-group where vasospasm was the sole contributory factor.

Amongst the 127 cases evaluated in total, mortality and severe morbidity on leaving the neurosurgical unit was 33% in the active group and 52% in the placebo group. This difference is significant (p = 0.05). The risk was reduced by 54% in the active group.

If mortality is considered alone, the risk is reduced by 80% as a result of treatment since mortality fell from 37% in group P to 9.6% in the active group (p = 0.001).

The decrease in mortality and severe morbidity is even more striking in the sub-group where the outcome is attributable to vasospasm alone (63% of cases evaluated – see above). Mortality and severe morbidity was 31.5% in group P compared with 11.0% in group N, representing a 73% reduction in risk with a significance of p = 0.01. The figures for mortality alone are 18.5% (P) and 2.7% (N). This represents an 84% reduction in risk with a significance of p = 0.01.

Discussion

This randomized study therefore confirms the findings made in the open study conducted by Koos et al.[13] concerning the efficacy of nimodipine in treating the ischaemic neurological consequences of established vasospasm. Koos et al. treated 91 patients already afflicted with neurological deficits and/or consciousness disturbances due to vasospasm following haemorrhage. Treatment consisted of i.v. doses of 24–48 mg nimodipine per day for 7–10 days, followed by oral treatment with 240 mg/day.

Treatment commenced within 24 hours of the onset of clinical manifestations of ischaemia in 67% of patients, within 24–48 hours in 12% and later than 48 hours after onset in 21%. 84% of patients were classified in Hunt and Hess grades III, IV or V at the start of treatment.

The outcome at the end of treatment was considered good in 65% and poor in 22% of cases. Overall mortality was 13%, while that attributable to vasospasm alone was 11%. Further analysis on the basis of the timing of treatment showed that the outcome was good in 69% of patients who underwent early treatment (within 24 hours) and only 53% in the group treated at a later stage (later than 48 hours).

The therapeutic activity of nimodipine on arterial spasm has also been documented following intracisternal administration of a solution of the drug during surgery[2]. Immediate dilatation of the artery of the order of 13 to 140% was observed[2]. Böker et al.[8] successfully relieved vasospasm following rupture of an aneurysm by intracarotid injection of nimodipine (0.2 mg/hour over 90 minutes). The increase in the calibre of the vessel was confirmed angiographically.

There thus seem to be firm grounds for believing that nimodipine has a therapeutic effect on established vasospasm, whether administered intravenously, intracisternally or injected into the carotid artery. Mortality and morbidity due to vasospasm are significantly reduced in patients treated intravenously. The earlier treatment is initiated, the greater the reduction. The risk was reduced by 73% in the French study in cases where vasospasm was the sole cause.

The fact remains that the activity of nimodipine is most in evidence in prophylaxis of vasospasm in terms of the extent of the reduction in the number of deaths and severe neurological consequences. It would appear on the basis of current knowledge that early surgery combined with prophylactic treatment of vasospasm with nimodipine represents the best option for patients, where feasible, following rupture of an aneurysm.

References

1. Allen GS, Ahn HS, Preziosi TJ et al (1983) Cerebral arterial spasm – a controlled trial of nimodipine in patients with subarachnoid hemorrhage. N Engl J Med 308: 619–624
2. Auer LM, Ito Z, Suzuki A, Ohta H (1982) Prevention of symptomatic vasospasm by topically applied nimodipine. Acta Neurochir (Wien) 63: 297–302
3. Auer LM (1983) Acute surgery of cerebral aneurysms and prevention of symptomatic vasospasm. Acta Neurochir (Wien) 69: 273–281
4. Auer LM (1984) Acute operation and preventive nimodipine improve outcome in patients with ruptured cerebral aneurysms. Neurosurgery 15: 57–66
5. Auer LM, Brandt L, Ebeling U et al (1986) Nimodipine and early aneurysm operation in good condition SAH patients. Acta Neurochir (Wien) 82: 7–13
6. Auer LM, Schneider GH, Auer Th (1986) Computerized tomography and prognosis in early aneurysms surgery. J Neurosurg 65: 217–221
7. Battye R (1985) Cooperative multi-institutional study of the safety and efficacy of nimodipine for the prevention and/or modification of ischemic neurological deficits due to cerebral arterial spasm following subarachnoid hemorrhage from ruptured intracranial aneurysms. Internal Report
8. Böker DK, Solymosi L, Wassmann H (1985) Immediate postangiographic intraarterial treatment of cerebral vasospasm after subarachnoid hemorrhage. Neurochirurgia 28: 118–120
9. Fisher CM, Roberson GH, Ojemann RG (1977) Cerebral vasospasm with ruptured saccular aneurysm – the clinical manifestation. Neurosurgery 1: 245–248
10. Fodstad H, Forssel A, Liliequist B, Schannong M (1981) Antifibrinolysis with tranexamic acid in aneurysmal subarachnoid hemorrhage. A consecutive controlled clinical trial. Neurosurgery 8: 158–165
11. Fodstad H, Liliequist B, Schannong M, Thulin CA (1978) Tranexamic acid in the preoperative management of ruptured intracranial aneurysms. Surg Neurol 9: 9–15
12. Kassel NF, Torner JC, Adams HP (1984) Antifibrinolytic therapy in the acute period following aneurysmal subarachnoid haemorrhage. J Neurosurg 61: 225–230
13. Koos WT, Perneczky A, Auer LM et al (1985) Nimodipine treatment of ischemic neurological deficits due to cerebral vasospasm after subarachnoid hemorrhage. Neurochirurgia 28: 114–117
14. Ljunggren B, Brandt L, Säveland H, Nilsson PE, Cronqvist S, Andersson KE, Vinge E (1984) Outcome in 60 consecutive patients treated with early aneurysm operation and intravenous nimodipine. J Neurosurg 61: 864–873
15. Ljunggren B, Säveland H, Brandt L (1983) Causes of unfavorable outcome after early aneurysm operation. Neurosurgery 13: 629–633
16. Mee EW, De Dorrance, Low D, Neil-Dwyer G (1986) Cerebral blood flow and neurological outcome: a controlled study of nimodipine in patients with subarachnoid haemorrhage. J Neurol Neurosurg Psychiatry 49: 469, full paper in preparation
17. Pasqualin A, Da Pian R (1982) An analysis of vasospasm following early surgery for intracranial aneurysms. Acta Neurochir (Wien) 63: 153–159
18. Philippon J, Grob R, Dagreon F, Guggiari M, Rivierez M, Viars P (1986) Prevention of vasospasm in subarachnoid haemorrhage. A controlled study with nimodipine. Acta Neurochir (Wien) 82: 110–114
19. Sahs AL, Perret GE, Locksley HB, Nishioka H (1969) Intracranial aneurysms and subarachnoid hemorrhage. A cooperative study. JB Lippincott Company, Philadelphia London
20. Säveland H, Ljunggren B, Brandt L, Messeter K (1986) Delayed ischemic deterioration in patients with early aneurysm operation and intravenous nimodipine. Neurosurgery 18: 146–150
21. Suzuki J, Kodama N, Yoshimoto, Mozoi K (1982) Ultraearly surgery of intracranial aneurysms. Acta Neurochir (Wien) 63: 186–191

22. Suzuki J, Komatsu S, Sato T, Sakurai Y (1980) Correlation between CT findings and subsequent development of cerebral infarction due to vasospasm in subarachnoid hemorrhage. Acta Neurochir (Wien) 56: 63–70

23. Taneda M (1982) Effect of early operation for ruptured aneurysms on prevention of delayed ischemic symptoms. J Neurosurg 57: 622–627

Correspondence: Dr. F. Buchheit, Neurosurgical Unit, Hautepierre Hospital, Strasbourg, France.

Neurological Research

A journal of progress in Neurosurgery and Neurosciences

Editor-in-chief: George Austin (Santa Barbara, Ca, USA)
Assistant Editor-in-chief: Bernard Pertuiset (Paris, France)

Neurological Research is an international journal for reporting both basic and clinical research in the fields of neurosurgery and neurosciences. It provides a medium for those who recognize the wider implications of their work and who wish to be informed of the relevant experience of others in related and more distant fields.

SCOPE

Neurological Research publishes original and fundamental studies on neurosurgery and related disciplines.

Regular features include research papers, review articles, short communications, book reviews and conference announcements.

For further details about the journal and a sample copy please contact Mrs Sheila King, Butterworth Scientific Ltd, P.O. Box 63, Westbury House, Bury Street, Guildford, Surrey GU2 5BH, UK.

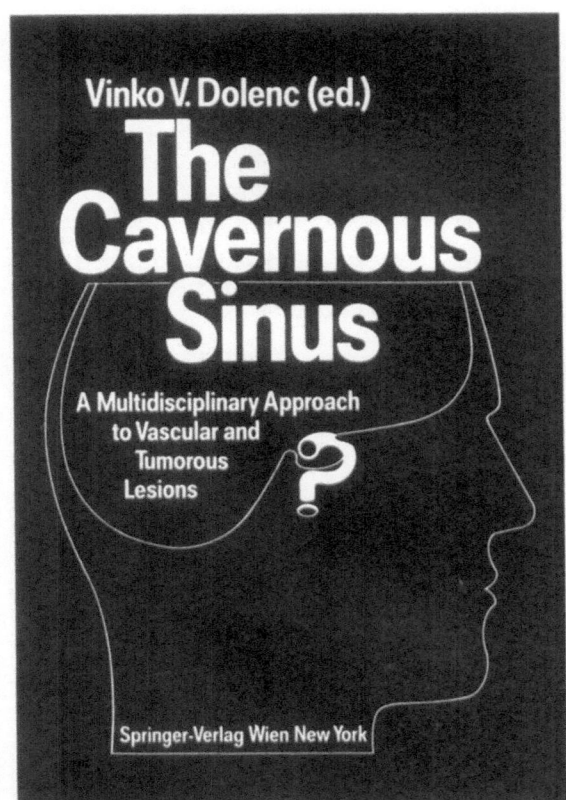